Springer Tracts in Advanced Robotics

Volume 35

Editors: Bruno Siciliano · Oussama Khatib · Frans Groen

Christian Laugier and Raja Chatila (Eds.)

Autonomous Navigation in Dynamic Environments

Springer

Professor Bruno Siciliano, Dipartimento di Informatica e Sistemistica, Universitá di Napoli Federico II, Via Claudio 21, 80125 Napoli, Italy, E-mail: siciliano@unina.it

Professor Oussama Khatib, Robotics Laboratory, Department of Computer Science, Stanford University, Stanford, CA 94305-9010, USA, E-mail: khatib@cs.stanford.edu

Professor Frans Groen, Department of Computer Science, Universiteit vanAmsterdam, Kruislaan 403, 1098 SJ Amsterdam, The Netherlands, E-mail: groen@science.uva.nl

Editors

Christian Laugier
INRIA Rhône-Alpes
655 Avenue de l'Europe
38334 Saint-Ismier Cedex
France
E-mail: Christian.Laugier@inrialpes.fr

Raja Chatila
LAAS-CNRS
7 Avenue Colonel Roche
31077 Toulouse Cedex 4
France
E-mail: Raja.Chatila@laas.fr

Library of Congress Control Number: 2007929544

ISSN print edition: 1610-7438
ISSN electronic edition: 1610-742X
ISBN-10 3-540-73421-X Springer Berlin Heidelberg New York
ISBN-13 978-3-540-73421-5 Springer Berlin Heidelberg New York

Springer is a part of Springer Science+Business Media
springer.com

Typesetting: Digital data supplied by editor.
Data-conversion and production: SPS, Chennai, India
Printed on acid-free paper SPIN: 11778271 89/SPS 5 4 3 2 1 0

Editorial Advisory Board

Preface

Autonomous navigation in open and dynamic environment is a challenging problem and an increasingly important topic for the near future applications of mobile robotics and intelligent vehicles. It is more and more obvious that future robot systems will have to share a common physical space with humans while intensively interacting with them, for the purpose of assisting peoples in various everyday tasks at work, at home, or during transportation. This means that such robots will clearly have to move safely in an open and unpredictable world, while interacting in various ways with peoples.

Navigation techniques in static environments are well known, but it is not clear how these techniques can cope with dynamic environments including people, other robots, and changing landmarks and environment features. Today robotics technologies have shown their ability to solve various navigation problems mainly in static environments; the success of the last DARPA Grand Challenge in USA is a clear example of such current robotics capabilities. However, moving to more dynamic spaces often populated by human is still an open issue. This is why the new DARPA Grand Challenge is focusing onto city environments and urban car traffic. From the application point of view, future Service Robots and Intelligent Vehicles are waiting for robust solutions to this problem.

The purpose of this book is to address the challenging problem of Autonomous Navigation in Dynamic Environments, and to present new ideas and approaches in this newly emerging technical domain. The book surveys the state-of-the-art, discusses in detail various related challenging technical aspects, and addresses upcoming technologies in this field. The aim of the book is to establish a foundation for a broad class of mobile robot mapping and navigation methodologies for indoor, outdoor, and exploratory missions.

Three main topics located on the cutting edge of the state of the art are addressed, from both the theoretical and technological point of views: Dynamic world understanding and modelling for safe navigation, Obstacle avoidance and motion planning in dynamic environments, and Human-robot physical interactions. Several models and approaches are proposed for solving problems such as Simultaneous Localization and Mapping (SLAM) in dynamic environments, Mobile obstacle detection and tracking, World state estimation and motion prediction, Safe navigation in dynamic environments, Motion planning in dynamic environments, Robust decision making under uncertainty, and Human-Robot physical interactions.

Christian Laugier
Research Director at INRIA (France)
Raja Chatila
Research Director at LAAS-CNRS (France)

Foreword

At the dawn of the new millennium, robotics is undergoing a major transformation in scope and dimension. From a largely dominant industrial focus, robotics is rapidly expanding into the challenges of unstructured environments. Interacting with, assisting, serving, and exploring with humans, the emerging robots will increasingly touch people and their lives.

The goal of the new series of Springer Tracts in Advanced Robotics (STAR) is to bring, in a timely fashion, the latest advances and developments in robotics on the basis of their significance and quality. It is our hope that the wider dissemination of research developments will stimulate more exchanges and collaborations among the research community and contribute to further advancement of this rapidly growing field.

The collection edited by Christian Laugier and Raja Chatila is the third one in the series on mapping and navigation, and is focused on the problem of autonomous navigation in dynamic environments. The state of the art is surveyed, a number of challenging technical aspects are discussed and upcoming technologies are addressed.

The ambitious goal is to lay down the foundation for a broad class of robot mapping and navigation methodologies for indoors, outdoors, and even exploratory missions. Future service robots and intelligent vehicles are waiting for effective solutions to such kind of problems.

The material is organised in three parts; namely, dynamic world understanding and modeling for safe navigation, obstacle avoidance and motion planning in dynamic environments, and human-robot interaction. Gathering some of the authorities working in the field, this volume constitutes a fine addition to the Series!

Naples, Italy
May 2007

Bruno Siciliano
STAR Editor

Contents

Part I
Dynamic World Understanding and Modelling for Safe Navigation

1

Mobile Robot Map Learning from Range Data in Dynamic Environments

Wolfram Burgard[1], Cyrill Stachniss[1,2], and Dirk Hähnel[1,3]

[1] University of Freiburg, Dept. of Computer Science, 79110 Freiburg, Germany
[2] Eidgenössische Technische Hochschule Zürich (ETH), 8092 Zürich, Switzerland
[3] Intel Research Seattle, 1100 NE 45th Street, Seattle, WA 98105, USA

Summary. The problem of generating maps with mobile robots has received considerable attention over the past years. Most of the techniques developed so far have been designed for situations in which the environment is static during the mapping process. Dynamic objects, however, can lead to serious errors in the resulting maps such as spurious objects or misalignments due to localization errors. In this chapter, we consider the problem of creating maps with mobile robots in dynamic environments. We present two approaches to deal with non-static objects. The first approach interleaves mapping and localization with a probabilistic technique to identify spurious measurements. Measurements corresponding to dynamic objects are then filtered out during the registration process. Additionally, we present an approach that learns typical configurations of dynamic areas in the environment of a mobile robot. Our approach clusters local grid maps to identify the typical configurations. This knowledge is then used to improve the localization capabilities of a mobile vehicle acting in dynamic environments. In practical experiments carried out with a mobile robot in a typical office environment, we demonstrate the advantages of our approaches.

1.1 Introduction

Learning maps with mobile robots is one of the fundamental problems in mobile robotics. In the literature, the mobile robot mapping problem is often referred to as the *simultaneous localization and mapping problem (SLAM)* [10, 13, 16, 20, 21, 31, 33]. This is because mapping includes both, estimating the position of the robot relative to the map and generating a map using the sensory input and the estimates about the robot's pose.

Whereas most of todays mapping systems are able to deal with noise in the odometry and noise in the sensor data, they assume that the environment is static during mapping. However, if a person walks through the sensor range of the robot during mapping, the resulting map will contain evidence about an object at the corresponding location. Moreover, if the robot scans the same area a second time and registers the two scans, the resulting pose estimates will be less accurate if the person has moved in between. Thus, dynamic objects can lead to spurious objects in the resulting maps and at the same time can make localization harder.

C. Laugier and R. Chatila (Eds.): Autonomous Navi. in Dyn. Environ., STAR 35, pp. 3–28, 2007.
springerlink.com

Throughout this chapter, we consider the problem of learning grid maps with mobile robots in dynamic environments. In principle, there are different ways to dealing with the dynamic aspects in an environment. The most naïve way of dealing with dynamic objects is to apply the standard map updating equations. Typical techniques in this context are the occupancy grid algorithm [25] or the counting model used throughout this chapter. Under both algorithms, areas assumed as occupied will certainly be regarded as free after the robot has seen them unoccupied for a long enough period of time (and vice versa). This is due to the fact that the map updating operations are additive as can be seen, for example, from the log-odds representation of occupancy grid maps [25] or directly from the counting model. Accordingly, the robot needs to see an area as free more often as it has seen it occupied to make it belief that the corresponding area is free. An alternative way is to identify whether or not a beam is reflected by a dynamic object. One popular way to achieve this is to use features corresponding to dynamic objects and to track such objects while the robot moves through its environment [17, 34]. Before the robot updates its map, it can then simply filter all measurements that correspond to dynamic objects. Whereas such approaches have been demonstrated to be quite robust, their disadvantage lies in the fact that the features need to be known a priori. Throughout this chapter, we will describe an alternative technique that applies the Expectation-Maximization (EM) algorithm. In the expectation step, we compute a probabilistic estimate about which measurements might correspond to static objects. In the maximization step, we use these estimates to determine the position of the robot and the map. This process is iterated until no further improvement can be achieved.

Whereas techniques for filtering dynamic aspects have been proven to be quite robust, their major disadvantage lies in the fact that the resulting maps only contain the static aspects of the environment. Throughout this chapter, we therefore will also describe an approach to explicitely model certain dynamic aspects of environments, namely the so-called low-dynamic or quasi-static states. Our approach is motivated by the fact, that many dynamic objects appear only in a limited number of possible configurations. As an example, consider the doors in an office environment, which are typically either open or closed. In such a situation, techniques to filter out dynamic objects produce maps which do not contain a single door. This can be problematic since in many corridor environments doors are important features for localization. The knowledge about the different possible configurations can explicitly improve the localization capabilities of a mobile robot. Therefore, it is important to integrate such information into the map of the environment.

The contribution of this chapter is novel approach to generating grid maps in dynamic environments from range data. Our algorithm first estimates for each individual beam whether or not it has been reflected by a dynamic object. It then uses this information during the range registration process to estimate get better pose estimates. It also learns the quasi-static states of areas by identifying sub-maps which have typical configurations. This is achieved by clustering local grid maps. We present experiments illustrating the accuracy of the resulting maps and also an extended Monte-Carlo localization algorithm, which uses the clusters of the local maps to more accurately localize the robot.

1.2 EM-Based Filtering of Beams Reflected by Dynamic Objects

As described above, one of the key problems in the context of mapping in dynamic environments is to determine whether or not a measurement is reflected by a dynamic object. Our approach to discover such measurements is strictly statistical. We use the popular EM-algorithm to identify data items that cannot be explained by the rest of the data set. The input to our routine is a sequence of data items $z = \{z_1, \ldots, z_T\}$. The output is a model m obtained from these data items after incorporating the estimates about spurious measurements. In essence, our approach seeks to identify a model m that maximizes the likelihood of the data. Throughout this chapter, we assume that each measurement z_t consists of multiple data $z_{t,1}, \ldots, z_{t,N}$ as it is the case, for example, for laser-range scans. Throughout this chapter, we assume that the data $z_{t,n}$ are beams obtained with a laser-range scanner.

To accurately map a dynamic environment, we need to know which measurements are caused by dynamic objects and therefore can safely be ignored in the alignment and map updating phase. To characterize spurious measurements in the data, we introduce additional variables $c_{t,n}$ that tell us for each t and each n whether the data item $z_{t,n}$ is caused by a static object or not. Each such variable $c_{t,n}$ is a binary variable that is either 0 or 1. It is 1 if and only if the $z_{t,n}$ is caused by a static object. The vector of all these variables will be denoted by c.

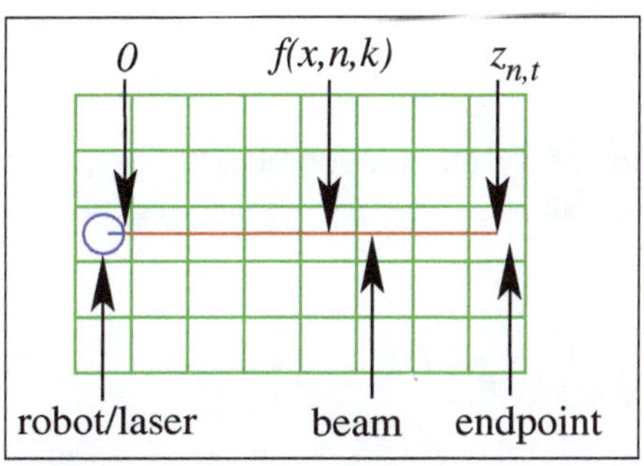

Fig. 1.1. Beam covering $z_{t,n}$ cells of a map

For the sake of simplicity, we give the derivation for beams that are parallel to the x-axis of the map. In this case, the length $z_{t,n}$ directly corresponds to the number of cells covered by this beam. We will later describe how to deal with beams that are not parallel to the x-axis. Let f be a function that returns for each position x_t of the robot, each beam number n, and each $k \leq z_{t,n}$ the index $f(x_t, n, k)$ of k-th field covered by that beam in the map (see Figure 1.1). To determine whether or not a beam is reflected by a dynamic object, we need to define the likelihood of a measurement given the current map m of the environment, the pose x of the robot, and the information about whether $z_{t,n}$ is reflected by a maximum range reading. Typically, maximum-range readings have to be treated differently, since those measurements generally are not reflected by any object. Throughout this chapter, we introduce indicator variables $\zeta_{t,n}$ which are 1 if and only if $z_{t,n}$ is a maximum range reading and 0, otherwise. The likelihood of a measurement $z_{t,n}$ given the value of $c_{t,n}$ and the map m can thus be computed as

$$\begin{aligned} p(z_{t,n} \mid c_{t,n}, x_t, m) &= \left[\prod_{k=0}^{z_{t,n}-1} (1 - m_{f(x_t,n,k)}) \right]^{\zeta_{t,n}} \\ &\cdot \left[[m_{f(x_t,n,z_{t,n})}]^{c_{t,n}} \cdot [1 - m_{f(x_t,n,z_{t,n})}]^{(1-c_{t,n})} \right. \\ &\cdot \left. \prod_{k=0}^{z_{t,n}-1} (1 - m_{f(x_t,n,k)}) \right]^{(1-\zeta_{t,n})} \end{aligned} \tag{1.1}$$

The first term of this equation specifies the likelihood of the beam given it is a maximum range scan. In such a situation, we compute the likelihood as the product of the probabilities that the beam has covered the cells 0 to $z_{t,n-1}$. Please note that the cell in which the beam ends does not provide any information since we do not know, whether there is an object or not given the beam is a maximum range reading. Thereby, the probability that a beam covers a cell $k < z_{t,n}$ is equal to $1 - m_{f(x_t,n,k)}$. The second row of this equation specifies how to deal with the case that a cell that reflects a non-maximum range beam. If $z_{t,n}$ is not reflected by a dynamic object, i.e. $c_{t,n} = 1$, then the likelihood equals $m_{f(x_t,n,z_{t,n})}$. If, in contrast, $z_{t,n}$ is reflected by a dynamic object, the likelihood is $1 - m_{f(x_t,n,z_{t,n})}$. As well as for the maximum range measurements, we have to consider in both cases that the beam has covered $z_{t,n} - 1$ cells before reaching cell $f(x_t, n, z_{t,n})$.

Based on the definition of the observation likelihood, we now will define the likelihood $p(z, c \mid x, m)$ of the data which we try to maximize in order to find the most likely map of the environment.

$$p(z, c \mid x, m) = \prod_{t=1}^{T} p(z_t, c_t \mid x_t, m) \tag{1.2}$$

$$= \prod_{t=1}^{T} p(z_t, \mid x_t, m) \cdot p(c_t \mid x_t, m) \tag{1.3}$$

$$= \prod_{t=1}^{T} p(z_t, \mid x_t, m) \cdot p(c_t) \tag{1.4}$$

$$= \prod_{t=1}^{T} \prod_{n=1}^{N} p(z_{t,n}, \mid c_{t,n}, x_t, m) \cdot p(c_t) \tag{1.5}$$

We obtain Equation (1.3) from Equation (1.2) by assuming that the z_t and c_t are independent given x_t and m. We furthermore consider c_t as independent from the location x_t and the map m, which leads to Equation (1.4). Finally, Equation (1.5) is derived from Equation (1.4) under the usual assumption, that the neighboring beams of a single scan are independent given the map of the environment.

Maximizing $p(z, c \mid x, m)$ is equivalent to maximizing the corresponding log likelihood, which can be derived from Equation (1.5) and Equation (1.1) by straightforward mathematical transformations.

$$\begin{aligned}
&\ln p(z, c \mid x, m) \\
&= \ln \prod_{t=1}^{T} \prod_{n=1}^{N} p(z_{t,n}, \mid c_{t,n}, x_t, m) \cdot p(c_t) \\
&= N \cdot \sum_{t=1}^{T} \ln p(c_t) + \sum_{t=1}^{T} \sum_{n=1}^{N} \ln p(z_{t,n}, \mid c_{t,n}, x_t, m) \\
&= N \cdot \sum_{t=1}^{T} \ln p(c_t) + \sum_{t=1}^{T} \sum_{n=1}^{N} \Bigg[(1 - \zeta_{t,n}) \cdot \Big[c_{t,n} \cdot \ln m_{f(x_t,n,z_{t,n})} \\
&\quad + (1 - c_{t,n}) \cdot \ln(1 - m_{f(x_t,n,z_{t,n})}) \Big] + \sum_{k=0}^{z_{t,n}-1} \ln(1 - m_{f(x_t,n,k)}) \Bigg]
\end{aligned} \tag{1.6}$$

Since the correspondence variables c are not observable in the first place, a common approach is to integrate over them, that is, to optimize the expected log likelihood $E_c[\ln p(c, z \mid x, m) \mid x, m, d]$ instead. Since the expectation is a linear operator, we can move it inside the expression. By exploiting the fact that the expectation of $c_{t,n}$ only depends on the corresponding measurement $z_{t,n}$ and the position x_t of the robot at that time. we can derive the following equation:

$$\begin{aligned}
&E_c[\ln p(z, c \mid x, m) \mid z, x, m] = \\
&\gamma + \sum_{t=1}^{T} \sum_{n=1}^{N} \Bigg[e_{t,n} \cdot (1 - \zeta_{t,n}) \cdot \ln m_{f(x_t,n,z_{t,n})} \\
&+ (1 - e_{t,n}) \cdot (1 - \zeta_{t,n}) \cdot \ln(1 - m_{f(x_t,n,z_{t,n})}) \\
&+ \sum_{k=0}^{z_{t,n}-1} \ln(1 - m_{f(x,n,k)}) \Bigg]
\end{aligned} \tag{1.7}$$

For the sake of brevity, we use the term

$$e_{t,n} = E_c[c_{t,n} \mid z_{t,n}, x_t, m] \tag{1.8}$$

in this equation. The term

$$\gamma = N \cdot \sum_{t=1}^{T} E_c[\ln p(c_t) \mid z, x, m] \tag{1.9}$$

is computed from the prior $p(c_t)$ of the measurements which is independent of z, x, and m. Accordingly, γ can be regarded as a constant.

Unfortunately, optimizing Equation (1.7) is not an easy endeavor. A typical approach to maximize log likelihoods is the EM algorithm. In the particular problem considered here, this amounts to generating a sequence of maps m of increasing likelihood. In the E-Step, we compute the expectations about the hidden variables c. In the M-step, we then compute the most likely map m using the expectations computed in the E-Step. Both steps are described in detail in the remainder of this section.

In the E-step, we compute the expectations $e_{t,n} = E_c[c_{t,n} \mid z_{t,n}, x_t, m]$ for each $c_{t,n}$ given the measurement $z_{t,n}$, the location x_t of the robot and the current map m. Exploiting the fact that $e_{t,n}$ equals $p(c_{t,n} \mid z_{t,n}, x_t, m)$ and considering the two cases that $z_{t,n}$ is a maximum range reading or not, we obtain:

$$e_{t,n} = \begin{cases} p(c_{t,n}) & \text{, if } \zeta_{t,n} = 1 \\ p(c_{t,n})\epsilon_{t,n} & \text{, otherwise} \end{cases}$$

where

$$\epsilon_{t,n} = \frac{1}{p(c_{t,n}) + (1 - p(c_{t,n}))(\frac{1}{m_{f(x_t,n,z_{t,n})}} - 1)} \tag{1.10}$$

The first equation corresponds to the situation that $z_{t,n}$ is a maximum range reading. Then, $e_{t,n}$ corresponds to the prior probability $p(c_{t,n})$ that a measurement is reflected by a static object. Thus, a maximum range reading does not provide any evidence about whether or not the cell in the map in which the beam ends is covered by a dynamic object.

In the M-Step, we want to determine the values for m and x that maximize Equation (1.7) after computing the expectations $e_{t,n}$ about the hidden variables $c_{t,n}$ in the E-step. Unfortunately, maximizing this equation is also not trivial since it involves a solution to a high-dimensional state estimation problem. To deal with the enormous complexity of the problem, many researchers phrase it as an incremental maximum likelihood process [33, 16]. The key idea of incremental approaches is to calculate the desired sequence of poses and the corresponding maps by maximizing the marginal likelihood of the t-th pose and map relative to the $(t-1)$-th pose and map. In our algorithm, we additionally consider the estimations $e_{t,n}$ that measurement n at time t is caused by a static object of the environment:

$$\hat{x}_t = \underset{x_t}{\operatorname{argmax}} \left\{ p(z_t \mid c_t, x_t, \hat{m}^{[t-1]}) \cdot p(x_t \mid u_{t-1}, \hat{x}_{t-1}) \right\} \tag{1.11}$$

In this equation, the term $p(z_t \mid c_t, x_t, \hat{m}^{[t-1]})$ is the likelihood of the measurement z_t given the pose $\hat{x}_t$ and the map $\hat{m}^{[t-1]}$ constructed so far. The term $p(x_t \mid u_{t-1}, \hat{x}_{t-1})$ represents the probability that the robot is at location x_t given the robot previously was at position $\hat{x}_{t-1}$ and has carried out (or measured) the motion u_{t-1}. The registration procedure is then carried out using the same algorithm as described in our previous work [17].

It remains to describe how the measurement z_t is then used to generate a new map $\hat{m}^{[t]}$ given the resulting pose $\hat{x}_t$ and the expectations $e_{t,n}$. Fortunately, once $x_1, \ldots, x_t$, have been computed, we can derive a closed-form solution for $m^{[t]}$. We want to determine the value of each field j of the map $m^{[t]}$ such that the overall likelihood of $m^{[t]}$

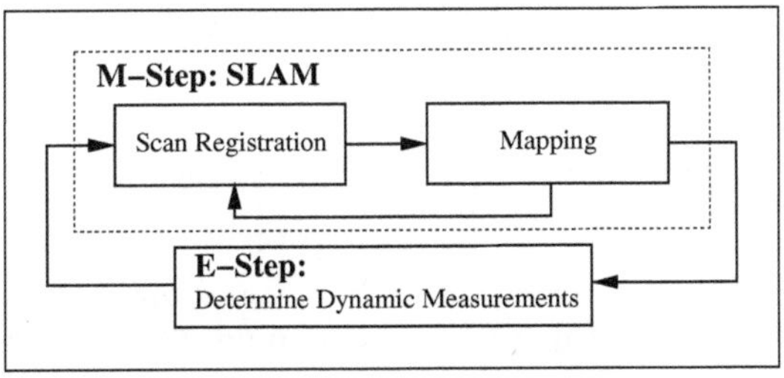

Fig. 1.2. Iteration of SLAM and dynamic beam estimation

is maximized. To achieve this, we sum over individual fields $j \in [1, \ldots, J]$ of the map. Thereby, we use an indicator function $I(y)$ which is 1, if y is true and 0, otherwise.

$$\hat{m}^{[t]} = \operatorname*{argmax}_{m} \left(\sum_{j=1}^{J} \sum_{t=1}^{T} \sum_{n=1}^{N} \left[I(f(x_t, n, z_{t,n}) = j) \right. \right.$$
$$\cdot (1 - \zeta_{t,n}) \cdot (e_{t,n} \ln m_j + (1 - e_{t,n}) \ln(1 - m_j))$$
$$\left. \left. + \sum_{k=0}^{z_{t,n}-1} I(f(x_t, n, k) = j) \cdot \ln(1 - m_j) \right] \right) \tag{1.12}$$

Now suppose, we define

$$\tilde{I}(x, n, k, j) := I(f(x, n, k) = j)$$

and

$$\alpha_j := \sum_{t=1}^{T} \sum_{n=1}^{N} \tilde{I}(x_t, n, z_{t,n}, j) \cdot (1 - \zeta_{t,n}) \cdot e_{t,n}$$
$$\beta_j := \sum_{t=1}^{T} \sum_{n=1}^{N} \left(\tilde{I}(x_t, n, z_{t,n}, j) \cdot (1 - \zeta_{t,n}) \right.$$
$$\left. \cdot (1 - e_{t,n}) + \sum_{k=0}^{z_{t,n}-1} I(f(x_t, n, k) = j) \right)$$

The quantity α_j corresponds to the sum of the expectations $e_{t,n}$ that beam n of scan t is reflected by a static object of all beams that are not maximum-range beams and that end in cell j. The term β_j, on the other hand, is the sum of two terms. The first term is the sum of the expectations $1 - e_{t,n}$ that beam n of scan t is reflected by a dynamic object of all beams that are not maximum-range beams and that end in cell j. The second value of the sum simply is the number of times a beam covers j but does not end in j. Please note that this value is independent from whether or not the corresponding beam is reflected by a dynamic object or not. Please furthermore note that in a static world with $e_{t,n} = 1$ for all t and n the term α_t corresponds to the number of times a beam that does not have the maximum length ends in j. In contrast to that, β_j is the number of times a beam covers a cell.

Using the definitions of α_j and β_j, Equation (1.12) turns into

$$m^{[t]} = \underset{m}{\operatorname{argmax}} \left(\sum_{j=1}^{J} \alpha_j \ln m_j + \beta_j \ln(1 - m_j) \right) \tag{1.13}$$

Since all m_j are independent, we maximize the overall sum by maximizing each m_j. A necessary condition to ensure that m_j is a maximum is that the first derivative equals zero:

$$\frac{\partial m}{\partial m_j} = \frac{\alpha_j}{m_j} - \frac{\beta_j}{1 - m_j} = 0 \tag{1.14}$$

By straightforward mathematical transformations we obtain

$$m_j = \frac{\alpha_j}{\alpha_j + \beta_j}. \tag{1.15}$$

Note that given the sensor model specified in Equation (1.1), this closed-form solution for the most likely map m for given positions x and static environments corresponds to the naïve counting technique. In this approach, one determines the probability that a map cell reflects a beam by counting how often a beam has ended in that cell and how often a beam has covered it without ending in it. This differs from the occupancy mapping approach in which one seeks to determine whether or not a particular area in the environment is occupied or not. To understand the difference between the two approaches, consider an object that reflects a beam in 70% of all cases. Whereas the counting model yields a value of 0.7 for this cell, the value of the same cell will typically converge to 1 in the context of occupancy grids.

The overall approach can be summarized as follows (see also Figure 1.2). We start with an initial map $\hat{m}$ obtained by the incremental mapping approach. Thereby, the expectations $e_{t,n}$ are initialized with the prior probability $p(c_{t,n})$ that a measurement is caused by a static object. Given the resulting map $\hat{m}$ and the corresponding positions $\hat{x}$, we compute new expectations $e_{t,n}$ for each beam according to Equation (1.8). These expectations are then used to compute a new map. The overall process is iterated until no improvement of the overall likelihood (Equation (1.6)) can be achieved or a certain number of iterations has been exceeded.

At the end of this section, we would like to discuss how to deal with beams that are not parallel to the x-axis. In this case, we no longer can compute the likelihood that a beam covers a cell j of m as $(1 - m_j)$. Otherwise, transversal beams covering more cells would accumulate a lower likelihood. The solution to this is to weigh the beams according to the length by which they cover a cell. Suppose B is the set of cells in m covered by a beam. Furthermore, suppose l_j is the length by which the beam covers field $j \in B$. Then, the likelihood of covering all cells in B is computed as $\prod_{j \in B} (1 - m_j)^{l_j}$.

This concludes the description of our algorithm for filtering measurements reflected from dynamic objects. As we will see in the experiments, this approach drastically improves the accuracy of the pose estimates and the quality of the resulting map.

1.3 Learning Maps of Quasi-static Environments

In certain situations, it can be advantageous to explicitely model dynamic aspects rather than simply filtering them out. As a motivating example consider the individual local maps depicted in Figure 1.3. These maps correspond to typical configurations of the same place and have been learned by a mobile robot operating in an office environment. They show a part of a corridor including two doors and their typical states. The approach described in this section learns such local configurations and to uses this information to improve the localization accuracy of a mobile robot.

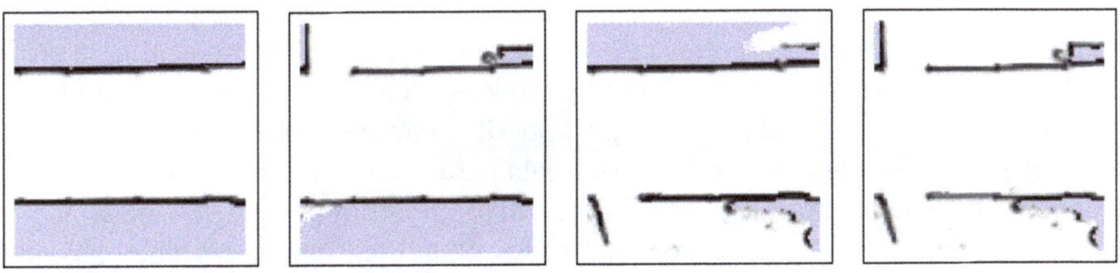

Fig. 1.3. Possible states of a local area. The different configurations correspond to open and closed doors.

The key idea of our approach is to use the information about changes in the environment during data acquisition to estimate possible spatial configurations and store them in the map model. To achieve this, we construct a sub-map for each area in which dynamic aspects have been observed. We then learn clusters of sub-maps that represent possible environmental states in the corresponding areas.

1.3.1 Map Segmentation

In general, the problem of learning maps in dynamic environments is a high-dimensional state estimation problem. A naïve approach could be to store an individual map of the whole environment for each potential state. Obviously, using this approach, one would have to store a number of maps that is exponential in the number of dynamic objects. In real world situations, the states of the objects in one room are typically independent of the states of the objects in another room. Therefore, it is reasonable to marginalize the local configurations of the individual objects.

Our algorithm segments the environment into local areas, called sub-maps. In this chapter, we use rectangular areas which inclose locally detected dynamic aspects to segment the environment into sub-maps. For each sub-map, the dynamic aspects are then modeled independently.

Note that in general the size of these local maps can vary from the size of the overall environment to the size of each grid cell. In the first case, we would have to deal with the exponential complexity mentioned above. In the second case, one heavily relies on the assumption that neighboring cells are independent, which is not justified in the context of dynamic objects. In our current system, we first identify positions in which the robot perceives contradictory observations which are typically caused by dynamic elements. Based on a region growing technique, areas which inclose dynamic aspects are determined. By taking into account visibility constraints between regions, they are merged until they do not exceed a maximum sub-map size (currently set to 20 m^2). This limits the number of dynamic objects per local map and in this way leads to a tractable complexity. An example for three sub-maps constructed in such a way is depicted in Figure 1.11. Note that each sub-map has an individual size and different sub-maps can (slightly) overlap.

1.3.2 Learning Environmental Configurations

To enable a robot to learn different states of the environment, we assume that the robot observes the same areas at different points in time. We cluster the local maps built from the different observations in order to extract possible configurations of the environment. To achieve this, we first segment the sensor data perceived by the robot into observation sequences. Whenever the robot leaves a sub-map, the current sequence ends and accordingly a new observation sequence starts as soon as the robot enters a new sub-map. Additionally, we start a new sequence whenever the robot moves through the same area for more than a certain amount of time (30 s). This results in a set Φ of observation sequences for each sub-map

$$\Phi = \{\phi_1, \ldots, \phi_n\}, \tag{1.16}$$

where each

$$\phi_i = z_{start(i)}, \ldots, z_{end(i)}. \tag{1.17}$$

Here z_t describes an observation obtained at time t. For each sequence ϕ_i of observations, we build an individual grid map for the corresponding local area. Thereby, we use the algorithm proposed in Section 1.2. Note that this approach eliminates highly dynamic aspects such as people walking by. Quasi-static aspects like doors, typically do not change their state frequently, so that the robot can observe them as static for the short time window. The different states are usually observed when the robot returns to a location at a later point in time.

Each grid computed for a local region is then transformed into a vector of probability values ranging from 0 to 1 and one additional value ξ to represent an unknown (unobserved) cell. All vectors which correspond to the same local area are clustered using the fuzzy k-means algorithm [14]. During clustering, we treat unknown cells in an slightly different way, since we do not want to get an extra cluster in case the sensor did not covered all parts of the local area. In our experiment, we obtained the best behavior using the following distance function for two vectors a and b during clustering

$$d(a, b) = \sum_i \begin{cases} (a_i - b_i) & a_i \neq \xi \wedge b_i \neq \xi \\ 0 & a_i = \xi \wedge b_i = \xi \\ \epsilon & \text{otherwise}, \end{cases} \tag{1.18}$$

where ϵ is a constant close to zero.

When comparing two values representing unknown cells, one in general should use the average distance computed over all known cells to estimate this quantity. In our experiments, we experienced that this leads to additional clusters in case a big part of a sub-map contains unknown cells even if the known areas of the maps were nearly identical. Therefore, we use the distance function given in Equation (1.18) which sets this distance value to zero.

Unfortunately, the number of different environmental states is not known in advance. Therefore, we iterate over the number of clusters and compute in each step a model using the fuzzy k-means algorithm. In each iteration, we create a new cluster initialized using the input vector which has the lowest likelihood under the current model. We evaluate each model θ using the Bayesian Information Criterion (BIC) [30]:

$$BIC = \log P(d \mid \theta) - \frac{|\theta|}{2} \log n \tag{1.19}$$

The BIC is a popular approach to score a model during clustering. It trades off the number $|\theta|$ of clusters in the model θ multiplied by the logarithm of the number of input vectors n and the quality of the model with respect to the given data d. The model with the highest BIC is chosen as the set of possible configurations, in the following also called patches, for that sub-map. This process is repeated for all sub-maps.

Note that our approach is an extension of the classical occupancy grid map [25] or counting model, in which the environment is not supposed to be static anymore. In situations without moving objects, the overall map reduces to a standard grid map.

The complexity of our mapping approach depends linearly on the number T of observations multiplied by the number s of sub-maps. Furthermore, the region growing applied to build up local maps introduces in the worst case a complexity of $p^2 \log p$, where p is the number of grid cells considered as dynamic. This leads to an overall complexity of $O(T \cdot s + p^2 \log p)$. Using a standard PC, our current implementation requires around 10% of the time needed to record the log file.

1.4 Monte-Carlo Localization Using Patch-Maps

It remains to describe how our patch-map representation can be used to estimate the pose of a mobile robot moving through its environment. Throughout this chapter, we apply an extension of Monte-Carlo localization (MCL), which has originally been developed for mobile robot localization in static environment [12]. MCL uses a set of weighted particles to represent possible poses of the robot. Typically, the state vector consists of the robot's position as well as its orientation. The sensor readings are used to compute the weight of each particle by estimating the likelihood of the observation given the pose of the particle and the map.

Besides the pose of the robot, we want to estimate the configuration of the environment in our approach. Since we do not use a static map like in standard MCL, we need to estimate the map $m^{[t]}$ as well as the pose x_t of the robot at time t

$$\begin{aligned} p(x_t, m^{[t]} \mid z_{1:t}, u_{0:t-1}) = \\ \eta \cdot p(z_t \mid x_t, m^{[t]}, z_{1:t-1}, u_{0:t-1}) \cdot p(x_t, m^{[t]} \mid z_{1:t-1}, u_{0:t-1}). \end{aligned} \tag{1.20}$$

Here η is a normalization constant and u_{t-1} refers to the motion command which guides the robot from x_{t-1} to x_t. The main difference to approaches on simultaneous localization and mapping (SLAM) is that we do not reason about all possible map configurations like SLAM approaches do. Our patch-map restricts the possible states according to the clustering of patches and therefore only a small number of configurations are possible.

Under the Markov assumption, the second line of Equation (1.20) can be transformed to

$$\begin{aligned}
& p(x_t, m^{[t]} \mid z_{1:t-1}, u_{0:t-1}) \\
&= \int_{x_{t-1}} \int_{m^{[t-1]}} p(x_t, m^{[t]} \mid x_{t-1}, m^{[t-1]}, z_{1:t-1}, u_{t-1}) \\
&\quad \cdot p(x_{t-1}, m^{[t-1]} \mid z_{1:t-1}, u_{0:t-2})\, dx_{t-1}\, dm^{[t-1]} \qquad (1.21) \\
&= \int_{x_{t-1}} \int_{m^{[t-1]}} p(x_t \mid x_{t-1}, m^{[t-1]}, z_{1:t-1}, u_{t-1}) \\
&\quad \cdot p(m^{[t]} \mid x_t, x_{t-1}, m^{[t-1]}, z_{1:t-1}, u_{t-1}) \\
&\quad \cdot p(x_{t-1}, m^{[t-1]} \mid z_{1:t-1}, u_{0:t-2})\, dx_{t-1}\, dm^{[t-1]} \qquad (1.22) \\
&= \int_{x_{t-1}} \int_{m^{[t-1]}} p(x_t \mid x_{t-1}, u_{t-1}) p(m^{[t]} \mid x_t, m^{[t-1]}) \\
&\quad \cdot p(x_{t-1}, m^{[t-1]} \mid z_{1:t-1}, u_{0:t-2})\, dx_{t-1}\, dm^{[t-1]}. \qquad (1.23)
\end{aligned}$$

Equation (1.23) is obtained from Equation (1.22) by assuming that $m^{[t]}$ is independent from $x_{t-1}, z_{1:t-1}, u_{t-1}$ given we know x_t and $m^{[t-1]}$ as well as assuming that x_t is independent from $m^{[t-1]}, z_{1:t-1}$ given we know x_{t-1} and u_{t-1}. Combining Equation (1.20) and Equation (1.23) leads to

$$\begin{aligned}
& p(x_t, m^{[t]} \mid z_{1:t}, u_{0:t-1}) \\
&= \eta \cdot p(z_t \mid x_t, m^{[t]}, z_{1:t-1}, u_{0:t-1}) \\
&\quad \int_{x_{t-1}} \int_{m^{[t-1]}} p(x_t \mid x_{t-1}, u_{t-1}) p(m^{[t]} \mid x_t, m^{[t-1]}) \\
&\quad \cdot p(x_{t-1}, m^{[t-1]} \mid z_{1:t-1}, u_{0:t-2})\, dx_{t-1}\, dm^{[t-1]}. \qquad (1.24)
\end{aligned}$$

Equation (1.24) describes how to extend the standard MCL approach so that it can deal with different environmental configurations. Besides the motion model $p(x_t \mid x_{t-1}, u_{t-1})$ of the robot, we need to specify a map transition model $p(m^{[t]} \mid x_t, m^{[t-1]})$, which describes the change in the environment over time.

In our current implementation, we do not reason about the state of the whole map, since each sub-map would introduce a new dimension in the state vector of each particle, which leads to a state estimation problem, that is exponential in the number of local sub-maps. Furthermore, the observations obtained with a mobile robot provide information only about the local environment of the robot. Therefore, we only estimate the state of the current patch the robot is in, which leads to one additional dimension in the state vector of the particles compared to standard MCL.

In principle, the map transition model $p(m^{[t]} \mid x_t, m^{[t-1]})$ can be learned while the robot moves through the environment. In our current system, we use a fixed density for all patches. We assume, that with probability α the current state of the environment does not change between time $t-1$ and t. Accordingly, the state changes to another configuration with probability $1-\alpha$. Whenever a particle stays in the same sub-map between $t-1$ and t, we draw a new local map configuration for that sample with probability $1-\alpha$. If a particle moves to a new sub-map, we draw the new map state from a uniform distribution over the possible patches in that sub-map. To improve the map transition model during localization, one in principle can update the values for α for each patch according to the observations of the robot. However, adapting these densities can also be problematic in case of a diverged filter or a multi-modal distribution about the pose of the robot. Therefore, we currently do not adapt the values of α while the robot acts in the environment.

Note that our representation bears resemblance with approaches using Rao-Blackwellized particle filters to solve the simultaneous localization and mapping problem [26, 23], as it separates the estimate about the pose of the robot from the estimate about the map. It computes the localization of the vehicle and uses this knowledge to identify the current state of the (local) map. The difference is that we aim to estimate the current state of the sub-map based on the possible configurations represented in our enhanced environmental model.

1.5 Experiments

The approaches described above has been implemented and tested on different robotic platforms, in different environments, and with 2d and 3d data acquired with SICK laser range finders. In all experiments, we figured out, that our approach can robustly filter out high-dynamic aspects. We present results demonstrating that the obtained maps contain fewer registration errors and less spurious objects. Additional experiments indicate that our approach can reliably model the quasi-static aspects of environments.

1.5.1 Filtering Dynamic Aspects

In the first set of experiments, we illustrate that our filtering algorithm can be used to reliably eliminate spurious measurements from range scans and at the same time reduces the registration errors.

Filtering People

The first experiments were carried out using a Pioneer I robot in Wean Hall of Carnegie Mellon University. There were several people walking through the environment while the robot was mapping it. The top image of Figure 1.4 shows the map obtained by a standard scan-matching procedure. As can be seen from the figure, the map contains many spurious objects and a huge number of registration errors. The most likely map resulting from the application of our approach is shown in the bottom image of Figure 1.4. The beams labeled as dynamic are drawn white in this figure. This demonstrates that our approach can reliably identify dynamic aspects and is able to learn maps that include the static aspects only.

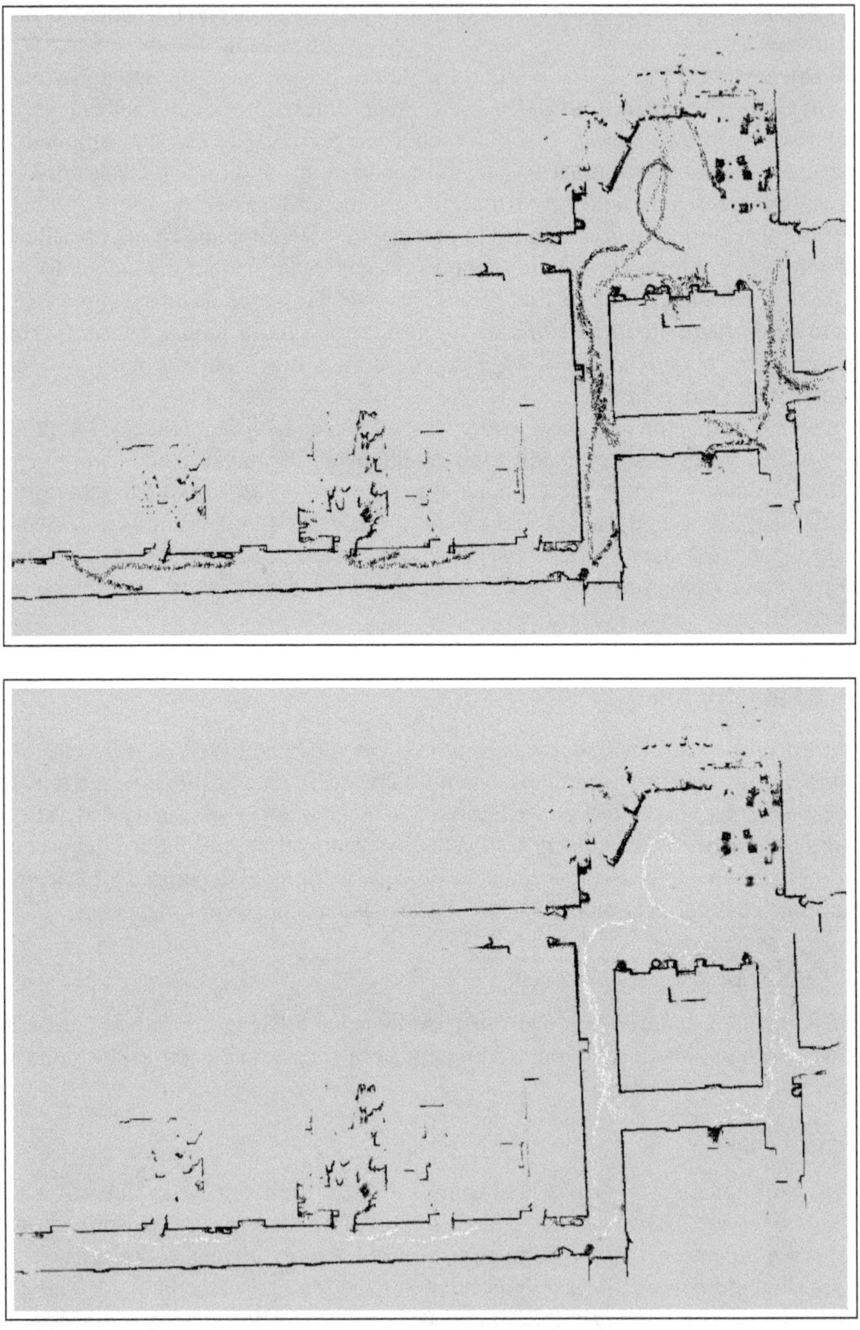

Fig. 1.4. Maps of Wean Hall at Carnegie Mellon University obtained without (top image) and with filtering measurements corrupted by dynamic objects (bottom image). The beams identified as reflected by dynamic objects are indicated by white dots.

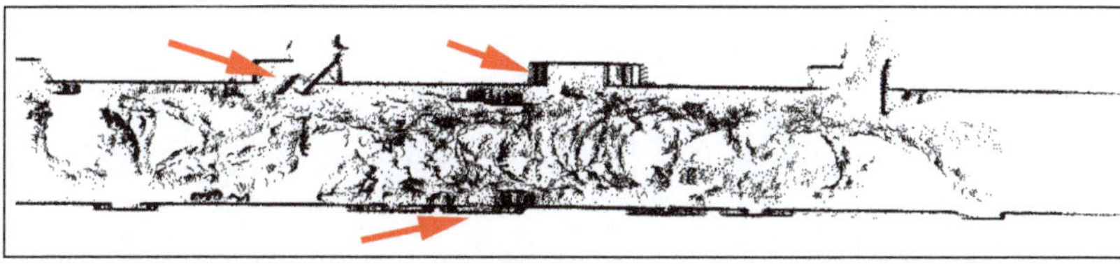

Fig. 1.5. Map obtained in a populated corridor of the Wean Hall at Carnegie Mellon University using the raw input data

Fig. 1.6. Map generated by our algorithm

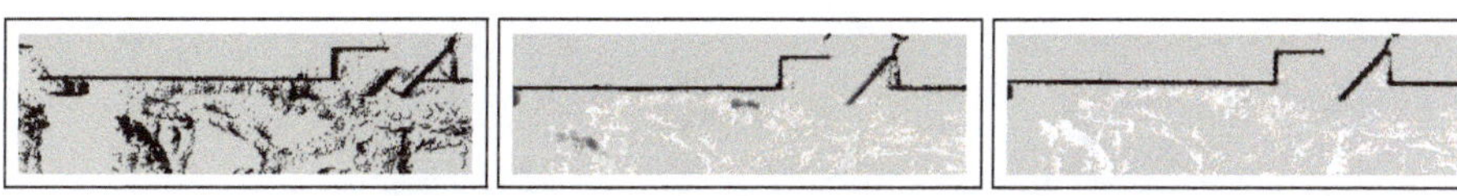

Fig. 1.7. Evolution of the map during EM. The images corresponds to iteration 1, 2, and 6.

Improved Registration Accuracy by Filtering Dynamic Objects

Besides the fact that the resulting maps contain less spurious objects, our approach also increases the localization accuracy. If dynamic objects are not handled appropriately during localization, matching errors become more likely. Figure 1.5 shows a typical map we obtained when mapping a densely populated environment. In this case, we mapped a part of the Wean Hall Corridor at Carnegie Mellon University during peak office hours when many persons were around. Some of them were trying to block the robot, so that the robot had to make detours around them. Therefore, the robot traveled $74\,\mathrm{m}$ with an average speed of $0.15\,\mathrm{m/s}$ ($0.35\,\mathrm{m/s}$ maximum). Despite the fact, that the huge amount of spurious objects make the map virtually useless for navigation tasks, the map also shows serious errors in the alignment. Some of the errors are indicated by arrows in the corresponding figure.

Figure 1.6 shows the map generated by our algorithm. As the figure illustrates, the spurious measurements (indicated by grey/orange dots) have been filtered out completely. Additionally, the alignment of the scans is more accurate.

Figure 1.7 depicts the evolution of a part of the map in the different rounds of the EM. It shows how the beams corresponding to dynamic objects slowly fade out and how the improved estimates about these beams improve the localization accuracy.

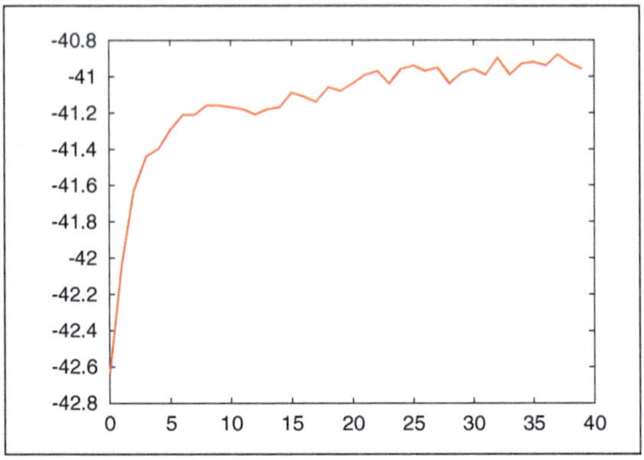

Fig. 1.8. Typical evolution of the log likelihood (Equation (1.6)) during the individual iterations of EM

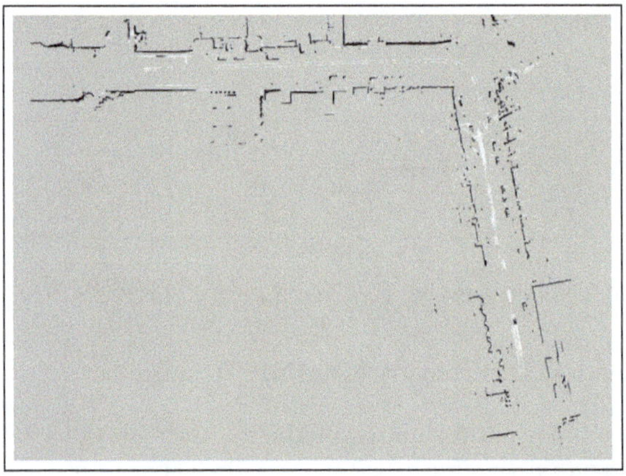

Fig. 1.9. Map of an outdoor scene after filtering dynamic objects

Figure 1.8 plots a typical evolution of $E_c[\ln p(c, z \mid x, m) \mid x, m, d]$ over the different iterations of our algorithm. It illustrates that our algorithm in fact maximizes the overall log likelihood. Please note that this curve generally is not monotonic because of the incremental maximum-likelihood solution to the SLAM problem. Slight variations in the pose can have negative effects in future steps, so that the map likelihood can decrease. However, we never observed significant decrease of the log likelihood.

Generating Large-Scale Outdoor Maps

To evaluate the capability of our technique to deal with arbitrary features, we mounted a laser-range scanner on a car and drove approximately 1 km through Pittsburgh, PA, USA (Corner between Craig Street and Forbes Avenue). The maximum speed of the car was 35 MPH in this experiment. We then applied our approach to the recorded data. The map generated by our algorithm is shown in Figure 1.9. Whereas the black dots correspond to the static objects in the scene, the white dots are those which are filtered out using our approach. Again, most of the dynamics of the scene could be removed. Only a few cars could not be identified as dynamic objects. This is mainly because we quickly passed cars waiting for turns and because we drove along the path only once. Please also note that due to the lack of a GPS, the map had to be computed without any odometry information.

Fig. 1.10. The images show textured 3d models of the Wean Hall lobby obtained without (left image) and with filtering (right image) dynamic aspects

Generating Textured 3D Maps

To demonstrate that our approach is not limited to 2d range data, we carried out several experiments with the mobile robot Robin which is equipped with a laser-scanner mounted on an AMTEC pan/tilt unit. On top of this scanner, we installed a camera which allows us to obtain textured 3d maps of an environment. Additionally, this robot contains a horizontally scanning laser range finder which we used in our experiments to determine dynamic objects. To label the beams in the 3d data as dynamic, we use a bounding box around the dynamic 2d points. To filter dynamic objects in the textures recorded with Robin's cameras, we choose for every polygon that image which has the highest likelihood of containing static aspects only. The left image of Figure 1.10 shows one particular view of a model obtained without filtering of dynamic objects. The arrow indicates a polygon whose texture contains fractions of an image of a person which walked through the scene while the robot was scanning it. After applying our approach, the corresponding beams and parts of the pictures were filtered out. The resulting model shown in the right image of Figure 1.10 therefore only contains textures showing static objects.

1.5.2 Learning Configurations of Environments

The second set of experiments is designed to illustrate that our approach to learning environmental configurations yields accurate models and at the same time improves the localization capabilities of a robot.

Application in an Office Environment

The first experiment on learning typical environmental configurations has been carried out in a typical office environment. The data was recorded by steering the robot through the environment while the states of the doors changed. To obtain a more accurate pose estimation than the raw odometry information, we apply an incremental scan-matching technique. Figure 1.11 depicts the resulting patch-map. For the three sub-maps that contain the doors whose states were changed during the experiment our algorithm was able to learn all configurations that occurred. The sub-maps and their corresponding patches are shown in the same figure.

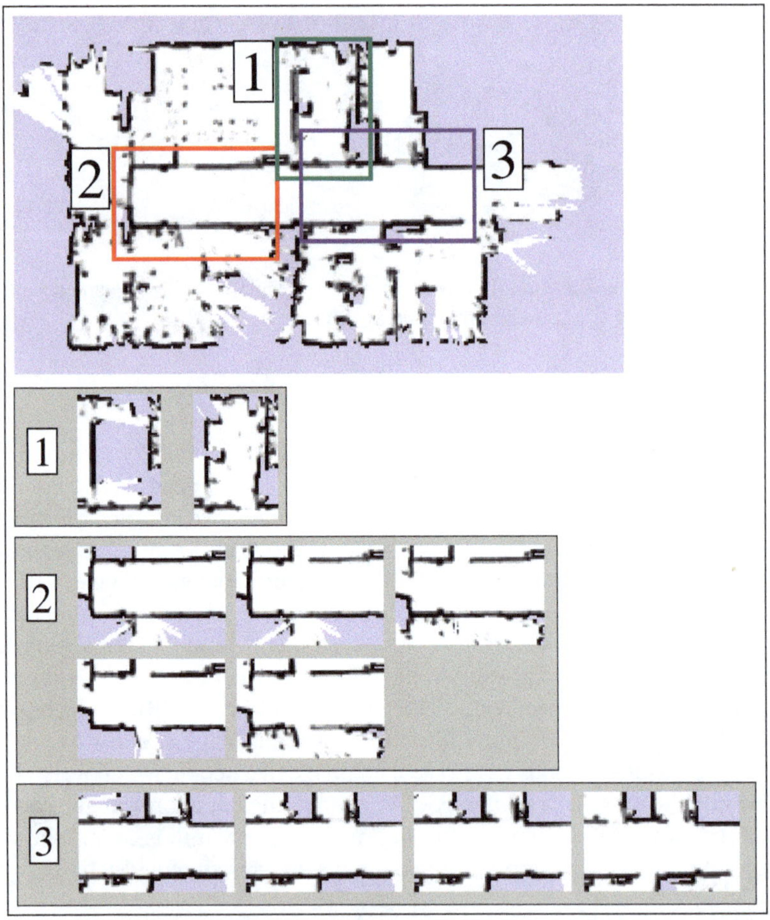

Fig. 1.11. A patch-map representing the different configurations learned for the individual sub-maps in a typical office environment

The second experiment is designed to illustrate the advantages of our map representation for mobile robot localization in quasi-static environments compared to standard MCL. The data used for this experiment was obtained in the same office environment as above. We placed a box at three different locations in the corridor. The resulting map including all patches obtained via clustering is depicted in Figure 1.12. Note that the

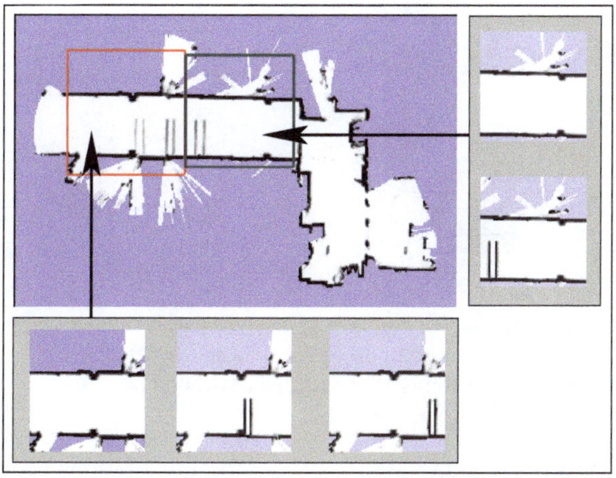

Fig. 1.12. A patch-map with the different configurations for the individual patches

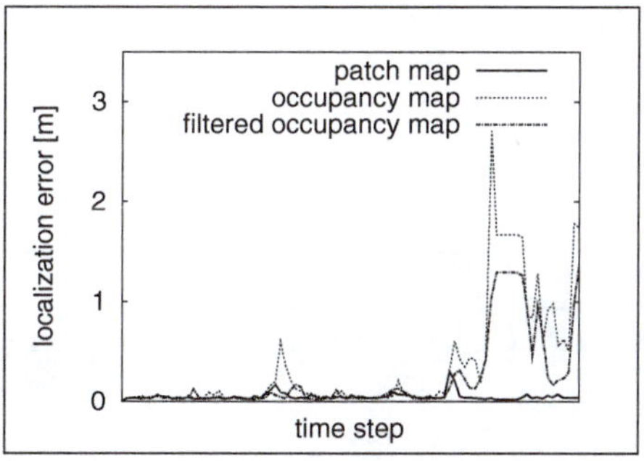

Fig. 1.13. The error in the pose estimate over time. As can be seen, using our approach the quality of the localization is higher compared to approaches using grid maps.

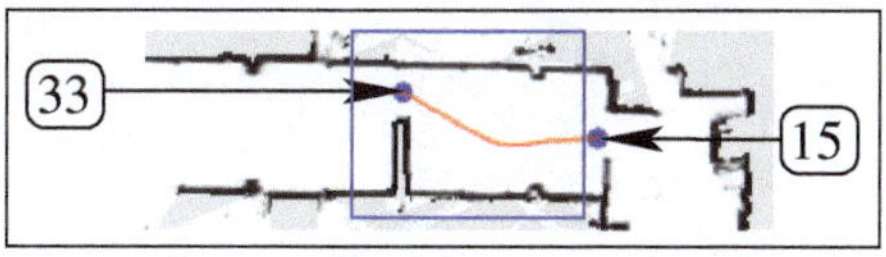

Fig. 1.14. The traveled path of the robot with two time labels. During its motion, the robot correctly identified the current state of the environment (see Figure 1.15).

tiles in the map illustrate the average over all patches. To evaluate the localization accuracy obtained with our map representation, we compare the pose estimates to that of a standard MCL using an occupancy grid map as well as a grid map obtained by filtering out dynamic objects [18].

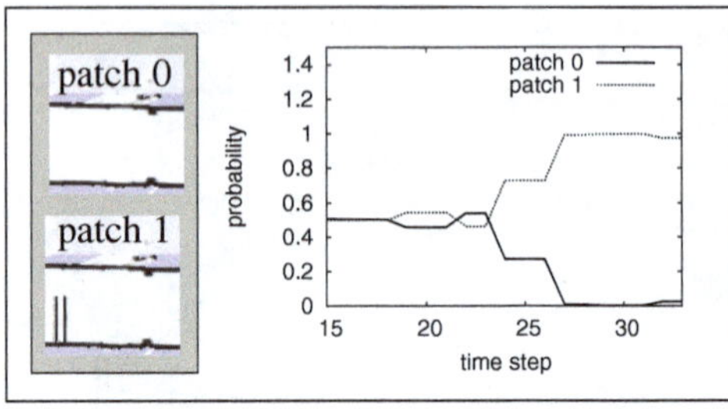

Fig. 1.15. The left image depicts the two possible patches whereas the graph on the right plots the probability of both patches according to the sample set. As can be seen, the robot identified that patch 1 correctly models the configuration of the environment.

Figure 1.13 plots the localization error over time for the three different representations. The error was determined as the weighted average distance from the poses of the particles to the ground truth, where each weight is given by the importance factor of the corresponding particle. In the beginning of this experiment, the robot traveled through static areas so that all localization methods performed equally well. Close to the end, the robot traveled through the dynamic areas, which results in high pose errors for both alternative approaches. In contrast to that, our technique constantly yields a high localization accuracy and correctly tracks the robot.

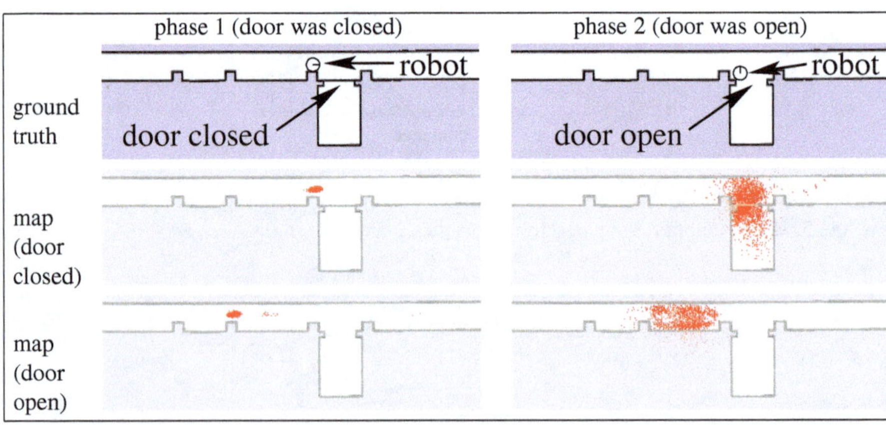

Fig. 1.16. In the beginning the door was closed (left column) but was later on opened (right column). The first row depicts the ground truth, whereas the second row illustrates the particle distributions in case the door is supposed to be closed in the occupancy grid map, whereas no door was mapped in the third row.

To further illustrate, how our extended MCL is able to estimate the current state of the environment, Figure 1.14 shows the path of the robot through a non-static area. Figure 1.15 plots the corresponding posterior probabilities for two different patches belonging to one sub-map. At time step 15, the robot entered the corresponding sub-map.

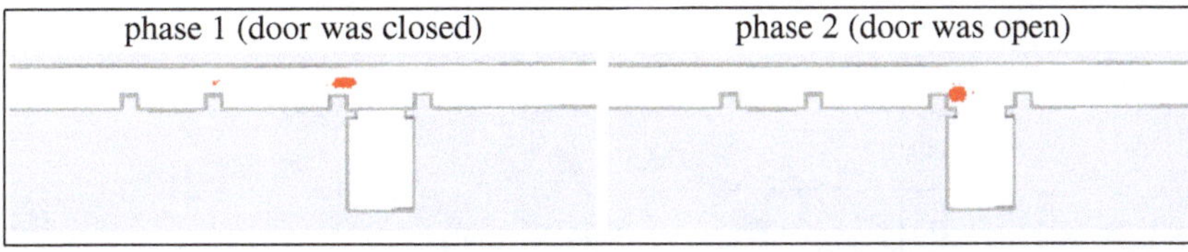

Fig. 1.17. Particle clouds obtained with our algorithm for the same situations as depicted in Figure 1.16

At this point in time, the robot correctly identified, that the particles, which localize the robot in patch 1, performed much better than the samples using patch 0. Due to the re-samplings in MCL, particles with a low importance weight are more likely to be replaced by particles with a high importance weight. Over a sequence of integrated measurements and re-samplings, this led to an probability close to 1 that the environment looked like the map represented by patch 1 (which exactly corresponded to the ground truth in that situation).

Global Localization

Additionally, we evaluated all three techniques in a simulated global localization task. We compared our approach using two patches to represent the state of the door with standard MCL using occupancy grid maps (see Figure 1.16 and 1.17). In one experiment, the occupancy grid map contained the closed door and in the second one the open door. During localization, the robot mostly moved in front of the door, which was closed in the beginning and opened in the second phase of the experiment.

As can be seen in left column of Figure 1.16 and 1.17, the MCL approach which uses the occupancy grid that models the closed door as well as our approach lead to a correct pose estimate. In contrast to that, the occupancy grid, which models the open door causes the filter to diverge. In the second phase of the experiment, the door was opened and the robot again moved some meters in front of the door (see right column of the same figure). At this point in time, the MCL technique using the occupancy grid, which models the closed door cannot track the correct pose anymore, whereas our approach is able to correctly estimate the pose of the robot. This simulated experiment again illustrates that the knowledge about possible configurations of the environment is important for mobile robot localization. Without this knowledge, the robot is not able to correctly estimate its pose in non-static environments.

Map Clustering

The last experiment is designed to illustrate the map clustering process. The input to the clustering was a set of 17 local grid maps. The fuzzy k-means clustering algorithm started with a single cluster, which is given by the mean computed over all 17 maps. The result is depicted in the first row of Figure 1.18. The algorithm then increased the number of clusters and re-computed the means in each step. In the fifth iteration the

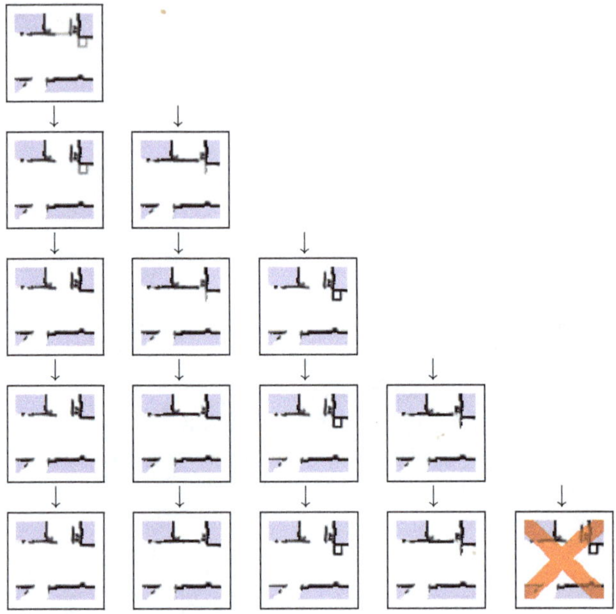

Fig. 1.18. Iterations of the map clustering process. Our approach repeatedly adds new clusters until no improvement is achieved by introducing new clusters (with respect to the BIC). Here, the algorithm ended up with 4 clusters, since cluster 3 and 5 are redundant.

newly created cluster is more or less equal to cluster 3. Therefore, the BIC decreased and the clustering algorithm terminated with the model depicted in the forth row of Figure 1.18.

1.6 Related Work

For mobile robots, there exist several approaches to mapping in dynamic environments. The approaches mostly relevant to the approach to filtering beams reflected by dynamic objects are the methods developed by Wang *et al.* [34] and our previous work described in [17]. Wang *et al.* [34] use a heuristic and feature-based approach to identify dynamic objects in range scans. The corresponding measurements are then filtered out during 2d scan registration. In our pervious work [17], we describe an approach to track persons in range scans and to remove the corresponding data during the registration and mapping process. Recently, Montesano *et al.* [24] describe an algorithm for simultaneously tracking moving objects and estimating the pose of the vehicle and landmarks. They also describe how to utilize the estimates during navigation. Compared to these techniques, our algorithm presented in this chapter does not rely on any pre-defined features or motion models. Rather, it considers every measurement individually and estimates a posterior about whether or not this data item has been generated by a dynamic object.

From a more general perspective, the problem of estimating dynamic aspects in data can be regarded as an outlier detection problem, since the spurious measurements are data items that do not correspond to the static aspects of the environment which are to be estimated. The identification of outliers is an important subtask in various application areas such as data mining [8, 19, 27], correspondence establishment [6, 11], clustering [14], or statistics [5]. In all these fields, errors in the data reduce the accuracy of the resulting models and thus can lead to a decreased performance whenever the learned models are used for prediction or robot navigation, for example. The problem considered in this chapter differs from these approaches in the fact that outliers cannot be detected solely based on their distance to the other data items. Rather, the measurements first have to be interpreted and transformed into a global representation (map) before individual measurements can be identified as outliers.

There has been work on updating maps or improving localization in populated environments. For example, in the system described in [9], a given static map is temporarily updated using the most recent sensory input. This allows the robot to consider areas blocked by people in the environment during path planning. Montemerlo *et al.* [22] present an approach to simultaneous localization and people tracking. This approach simultaneously maintains multiple hypotheses about the pose of the robot and people in its vicinity and in this way yields more robust estimates. Siegwart *et al.* [32] present a team of tour-guide robots that operates in a populated exhibition. Their system uses line features for localization and has been reported to successfully filter range-measurements reflected by persons. Fox *et al.* [15] present a probabilistic technique to identify range measurements that do not correspond to the given model of the environment. In contrast to our approach, these methods require a given and fixed map which is used for localization and for the extraction of the features corresponding to the people. Our filtering technique, in contrast, does not require a given map. Rather it learns the map from scratch using the data acquired with the robot's sensors.

Additionally, several authors have proposed techniques for learning dynamic objects of maps of dynamic aspects with mobile robots. For example, Anguelov *et al.* [2] present an approach which aims to learn models of non-stationary objects from proximity data. The object shapes are estimated by applying a hierarchical EM algorithm based on occupancy grids recorded at different points in time. The main difference to our approach to model quasi-static aspects is that we estimate typical configurations of the environment and do not address the problem of learning geometric models for different types of non-stationary obstacles.

Schulz and Burgard [29] proposed a probabilistic algorithm to estimate the state of dynamic objects in an environment. Avots *et al.* [4] apply a Rao-Blackwellized particle filter to estimate the state of doors. Both approaches, however, require a parameterized model of the environment that includes the dynamic objects such as doors. Anguelov *et al.* [3] uses an EM-based approach to detect doors and to estimate their states. Thereby, they take into account shape, color, and motion properties of wall-like objects. In contrast to these works, the approach presented in this chapter is independent of the type of quasi-static object and can learn arbitrary configurations of the environment.

Yamauchi and Beer [35] describe a network of places in which links model a connection between different places. These links may dynamically change their traversability. To deal with these dynamic aspects, they store a confidence value which is updated according to successful or unsuccessful attempts to traverses that link. In the context of landmark-based mapping, the approach presented by Andrade-Cetto and Senafeliu [1] is able to remove landmarks which are not observed anymore from the posterior.

Romero *et al.* [28] describe an approach to globally localize a mobile robot in static environments in which a clustering algorithm is applied to group similar places in the environment. In this way, the robot is able to reduce the number of possible pose hypotheses which speeds up the probabilistic localization process.

In a very recent work, Bieber and Duckett [7] proposed an approach that incorporates changes of the environment into the map representation. Compared to our work, they model temporal changes of local maps whereas we aim to identify the different configurations of the environment.

Our approach to learning typical configurations is designed to explicitely model possible states of the environment, like, e.g., open and closed doors or moved tables. As we have demonstrated in this chapter, it can be used in addition to the filtering techniques. We also demonstrated that the different environmental state hypotheses enable a mobile robot to more reliably localize itself and to also estimate the current configuration of its surroundings.

1.7 Conclusions

In this chapter, we presented probabilistic approaches to mapping in dynamic environments. We first presented an algorithm based on the EM algorithm that interleaves the identification of measurements that correspond to dynamic objects with a mapping and localization algorithm. In this way, it incrementally improves its estimate about spurious measurements and the quality of the map. The finally obtained maps contain less spurious objects and also are more accurate because of the improved range registration. Additionally, we presented a novel approach to model quasi-static environments using a mobile robot. In areas where dynamic aspects are detected, our approach creates local maps and estimates for each sub-map clusters of possible configurations of the corresponding space in the environment. Furthermore, we described how to extend Monte-Carlo localization to utilize the information about the different possible environmental states while localizing a vehicle. Our approach has been implemented and tested on real robots as well as in simulation. The experiments demonstrate, that our technique yields a higher localization accuracy compared to Monte-Carlo localization based on standard grid maps even such obtained after filtering out measurements reflected by dynamic objects.

Our techniques have been implemented and tested on different platforms. In several experiments carried out in indoor and outdoor environments we demonstrated that our approaches yield highly accurate maps. The results illustrate that our approaches can reliably estimate filter beams reflected by dynamic environments and that quasi-static aspects can be modeled accurately.

Acknowledgment

The authors would like to thank Sebastian Thrun and Rudolph Triebel for fruitful discussions. This work has partly been supported by the German Research Foundation (DFG) under contract number SFB/TR-8 (project A3) and by the EC under contract number FP6-004250-CoSy and FP6-IST-027140-BACS.

References

1. J. Andrade-Cetto and A. Sanfeliu. Concurrent map building and localization in indoor dynamic environments. *Int. Journal of Pattern Recognition and Artificial Intelligence*, 16(3):361–274, 2002.
2. D. Anguelov, R. Biswas, D. Koller, B. Limketkai, S. Sanner, and S. Thrun. Learning hierarchical object maps of non-stationary environments with mobile robots. In *Proc. of the Conf. on Uncertainty in Artificial Intelligence (UAI)*, 2002.
3. D. Anguelov, D. Koller, E. Parker, and S. Thrun. Detecting and modeling doors with mobile robots. In *Proc. of the IEEE Int. Conf. on Robotics & Automation (ICRA)*, pages 3777–3774, New Orleans, LA, USA, 2004.
4. D. Avots, E. Lim, R. Thibaux, and S. Thrun. A probabilistic technique for simultaneous localization and door state estimation with mobile robots in dynamic environments. In *Proc. of the IEEE/RSJ Int. Conf. on Intelligent Robots and Systems (IROS)*, pages 521–526, Lausanne, Switzerland, 2002.
5. V. Barnett and T. Lewis. *Outliers in Statistical Data.* Wiley, New York, 1994.
6. P. Besl and N. McKay. A method for registration of 3d shapes. *Trans. Patt. Anal. Mach. Intell. 14(2)*, pages 239–256, 1992.
7. P. Biber and T. Duckett. Dynamic maps for long-term operation of mobile service robots. In *Proc. of Robotics: Science and Systems (RSS)*, 2005. To appear.
8. C.E. Brodley and M.A. Friedl. Identifying and eliminating mislabeled training instances. In *Proc. of the National Conference on Artificial Intelligence (AAAI)*, 1996.
9. W. Burgard, A.B. Cremers, D. Fox, D. Hähnel, G. Lakemeyer, D. Schulz, W. Steiner, and S. Thrun. Experiences with an interactive museum tour-guide robot. *Artificial Intelligence*, 114(1-2), 2000.
10. J.A. Castellanos, J.M.M. Montiel, J. Neira, and J.D. Tardós. The SPmap: A probabilistic framework for simultaneous localization and map building. *IEEE Transactions on Robotics and Automation*, 15(5):948–953, 1999.
11. I.J. Cox and S.L. Hingorani. An efficient implementation of reid's multiple hypothesis tracking algorithm and its evaluation for the purpose of visual tracking. *IEEE Transactions on PAMI*, 18(2):138–150, February 1996.
12. F. Dellaert, D. Fox, W. Burgard, and S. Thrun. Monte carlo localization for mobile robots. In *Proc. of the IEEE Int. Conf. on Robotics & Automation (ICRA)*, Leuven, Belgium, 1998.
13. G. Dissanayake, H. Durrant-Whyte, and T. Bailey. A computationally efficient solution to the simultaneous localisation and map building (SLAM) problem. In *ICRA'2000 Workshop on Mobile Robot Navigation and Mapping*, San Francisco, CA, USA, 2000.
14. R. Duda, P. Hart, and D. Stork. *Pattern Classification*. Wiley-Interscience, 2001.
15. D. Fox, W. Burgard, and S. Thrun. Markov localization for mobile robots in dynamic environments. *Journal of Artificial Intelligence Research (JAIR)*, 11:391–427, 1999.
16. J.-S. Gutmann and K. Konolige. Incremental mapping of large cyclic environments. In *Proc. of the IEEE Int. Symp. on Computational Intelligence in Robotics and Automation (CIRA)*, 1999.
17. D. Hähnel, D. Schulz, and W. Burgard. Mobile robot mapping in populated environments. *Journal of the Robotics Society of Japan (JRSJ)*, 7(17):579–598, 2003.
18. D. Hähnel, R. Triebel, W. Burgard, and S. Thrun. Map building with mobile robots in dynamic environments. In *Proc. of the IEEE Int. Conf. on Robotics & Automation (ICRA)*, 2003.
19. George H. John. Robust decision trees: Removing outliers from databases. In *First International Conference on Knowledge Discovery and Data Mining*, pages 174–179, 1995".

20. J.J. Leonard and H.J.S. Feder. A computationally efficient method for large-scale concurrent mapping and localization. In J. Hollerbach and D. Koditschek, editors, *Proceedings of the Ninth International Symposium on Robotics Research*, pages 169–179, 2000.
21. F. Lu and E. Milios. Globally consistent range scan alignment for environment mapping. *Autonomous Robots*, 4:333–349, 1997.
22. M. Montemerlo and S. Thrun. Conditional particle filters for simultaneous mobile robot localization and people-tracking (slap). In *Proc. of the IEEE Int. Conf. on Robotics & Automation (ICRA)*, 2002.
23. M. Montemerlo, S. Thrun, D. Koller, and B. Wegbreit. FastSLAM: A factored solution to simultaneous localization and mapping. In *Proc. of the National Conference on Artificial Intelligence (AAAI)*, Edmonton, Canada, 2002.
24. L. Montesano, J. Minguez, and L. Montano. Modeling the static and the dynamic parts of the environment to improve sensor-based navigation. In *Proc. of the IEEE Int. Conf. on Robotics & Automation (ICRA)*, 2005.
25. H.P. Moravec and A.E. Elfes. High resolution maps from wide angle sonar. In *Proc. of the IEEE Int. Conf. on Robotics & Automation (ICRA)*, pages 116–121, St. Louis, MO, USA, 1985.
26. K. Murphy. Bayesian map learning in dynamic environments. In *Neural Info. Proc. Systems (NIPS)*, Denver, CO, USA, 1999.
27. S. Ramaswamy, R. Rastogi, and S. Kyuseok. Efficient algorithms for mining outliers from large data sets. In *Proc. of the ACM SIGMOD International Conference on Management of Data*, 2000.
28. L. Romero, E. Morales, and E. Sucar. A hybrid approach to solve the global localozation problem for indoor mobile robots considering sensor's perceptual limitations. In *Proc. of the Int. Conf. on Artificial Intelligence (IJCAI)*, 2001.
29. D. Schulz and W. Burgard. Probabilistic state estimation of dynamic objects with a moving mobile robot. *Robotics and Autonomous Systems*, 34(2-3):107–115, 2001.
30. G. Schwarz. Estimating the dimension of a model. *The Annals of Statistics*, 6(2), 1978.
31. H. Shatkay. *Learning Models for Robot Navigation*. PhD thesis, Computer Science Department, Brown University, Providence, RI, 1998.
32. R. Siegwart, K.O. Arras, S. Bouabdallah, D. Burnier, G. Froidevaux, X. Greppin, B. Jensen, A. Lorotte, L. Mayor, M. Meisser, R. Philippsen, R. Piguet, G. Ramel, G. Terrien, and N. Tomatis. Robox at Expo.02: A large-scale installation of personal robots. *Journal of Robotics & Autonomous Systems*, 42(3-4), 2003.
33. S. Thrun. A probabilistic online mapping algorithm for teams of mobile robots. *Int. Journal of Robotics Research*, 20(5):335–363, 2001.
34. C.-C. Wang and C. Thorpe. Simultaneous localization and mapping with detection and tracking of moving objects. In *Proc. of the IEEE Int. Conf. on Robotics & Automation (ICRA)*, 2002.
35. B. Yamauchi and R. Beer. Spatial learning for navigation in dynamic environments. *IEEE Transactions on Systems, Man and Cybernetics*, 26(3):496–505, 1996.

2

Optical Flow Approaches for Self-supervised Learning in Autonomous Mobile Robot Navigation

Andrew Lookingbill, David Lieb, and Sebastian Thrun

Stanford Artificial Intelligence Laboratory
Stanford University
Gates Hall 1A; 355 Serra Mall; Stanford, CA 94305-9010
{apml,dflieb,thrun}@stanford.edu

Summary. A common theme in autonomous mobile robotics is the desire to sense farther ahead of the robot than current approaches allow. This greater range would enable earlier recognition of hazards, better path planning, and higher speeds. In scenarios where the long range sensor modality is computer vision this has led to interest in developing techniques that can effectively identify and respond to obstacles at greater distances than those for which stereo vision methods are useful. This paper presents work on optical flow techniques that leverage the difference in appearance between objects at close range and the same objects at more distant locations in order to interpret monocular video streams in a useful manner. In particular, two applications are discussed: self-supervised off-road autonomous navigation, and adaptive road following in unstructured environments. Examples of the utility of the optical flow techniques discussed here in both arenas are provided.

2.1 Introduction

Recent years have witnessed significant growth in the field of autonomous mobile robot navigation. The varied applications of this technology have resulted in an increased research interest in this area. This interest is exemplified by competitions such as the DARPA Learning Applied to Ground Robots (LAGR) program [1].

Most current mobile robotic systems rely on 3D sensors such as laser rangefinders or stereo vision as their primary means of long range perception [2][3][4][5]. While providing accurate readings within their range of operation, these types of sensing modalities limit the platform's range of perception, and therefore its theoretical top speed. One solution to this problem is the use of monocular cameras for perception, since cameras typically capture scene information all the way to the horizon. Work in this area includes the use of color information to distinguish between pixels corresponding to the ground and those assumed to correspond to obstacles either because of dissimilarity in color [6] or spatial color gradient changes [7]. Texture is also a powerful monocular cue for obstacle detection. The use of texture has evolved from being used by MIT's Polly tour guide to differentiate between the floor and obstacles placed on the floor in an

C. Laugier and R. Chatila (Eds.): Autonomous Navi. in Dyn. Environ., STAR 35, pp. 29–44, 2007.
springerlink.com

indoor environment [8], to natural terrain classifiers suitable for use on planetary rovers [9].

However, due to the perspective nature of cameras, the sensor resolution per area decreases as distance from the robot increases. In order to make navigation decisions about distant objects based on camera images, the robot would need to infer information from a small patch of the image plane. Additionally, objects perceived at great distance can often change substantially in appearance in the image plane as the robot moves closer.

Another technique used successfully in autonomous robotics for obstacle avoidance is the use of optical flow. Image flow divergence can be used to navigate indoor hallways by estimating time to collision [10] while differences in the magnitude of optical flow can be used to segment objects moving at different speeds to aid in road navigation [11]. This paper discusses another use of optical flow: correlating the appearance of objects at a large distance from a robot with objects directly in front of the robot.

The two problems of image-based navigation, namely limited resolution for distant objects and the changing nature of the size and appearance of objects with changing distance, lend themselves to the use of learning techniques. By learning correlations between the appearance of distant objects and their traversibility characteristics, autonomous robots can make navigation decisions based on image patches consisting of a very small number of pixels. Using the local sensors present on most mobile robots (bumpers, infrared sensors, wheel encoders, stereo cameras, etc), accurate traversibility characteristics can be measured and associated with distant objects through a self-supervised learning procedure.

In this paper, we present a self-supervised learning approach using optical flow measurements to perform the association of local sensor readings to distant image pixels. To demonstrate the effectiveness of our approach, we present two applications: obstacle avoidance for an off-road autonomous robot and road detection in unstructured environments. In both applications, the frame-to-frame optical flow field is calculated and cached, permitting nearby regions to be associated with their appearance at greater distances.

2.2 Self-supervised Learning

Reverse Optical Flow: Our approach uses standard optical flow procedures to compile a history of inter-frame optical flow as a robot navigates through its surroundings. This reverse optical flow can be used to trace any object in the current frame back to its position in a previous frame. This allows the visual characteristics of the object to be examined at both near and far ranges.

In order to find this reverse optical flow chain, the optical flow between adjacent video frames must be calculated for a sequence of frames up to the current frame. For each pair of consecutive frames, a set of easily trackable, unique features are found in the first frame and then are traced back to their locations in the preceding frame. The displacement vectors between these two sets of positions comprise the optical flow field. In our approach, features are initially identified using the

Fig. 2.1. White lines show the optical flow vectors between two successive images in a video sequence

Shi-Tomasi algorithm [12]. This algorithm finds unambiguous features by finding regions in the frame containing a significant spatial image gradient in two orthogonal directions. Feature tracking between two consecutive frames is then carried out using a pyramidal implementation of the Lucas-Kanade tracker [13]. This tracker constructs image pyramids of filtered and subsampled versions of the original images. The displacement vectors between the feature locations in the two frames are calculated by iteratively maximizing a correlation measure over a small window, from the coarsest level down to the original level. The optical flow field between two consecutive video frames from a data set taken in the California Mojave Desert is illustrated by small lines in Fig. 2.1.

By subdividing the optical flow field into a rectangular grid, it can be subsampled and compressed by saving only the mean displacement vector within a cell as is shown in Fig. 2.2. In this representation, the optical flow fields for a large number of consecutive frames can be stored and quickly accessed in an array structure. In this way, a point in the current video frame can be followed back to its origin in any previous frame for which the optical flow has been stored by summing the displacement vectors. Fig. 2.3a shows a number of points that have been chosen in a single frame of input video, in Fig. 2.3b these same points have been traced back to their original locations 200 frames in the past.

The two applications discussed in this paper use the information from the reverse optical flow in two different ways. In the case of offroad navigation, the reverse optical flow is used to correlate the appearance of obstacles that have been identified using the short-range sensors of the robot with the appearance of similar obstacles at greater distances from the robot. This information is then used to segment the current video frame into hazardous and non-hazardous regions. This segmented image is then fed to a simple image space controller. The process is shown in Fig. 2.4. In the case of adaptive road following, the reverse optical flow information is used as follows: The region on which the vehicle currently sits is used as the definition of the roadway, and the algorithm subsequently follows regions matching this description. The most recent characteristics of the road are learned by examining the appearance of the current vehicle location in

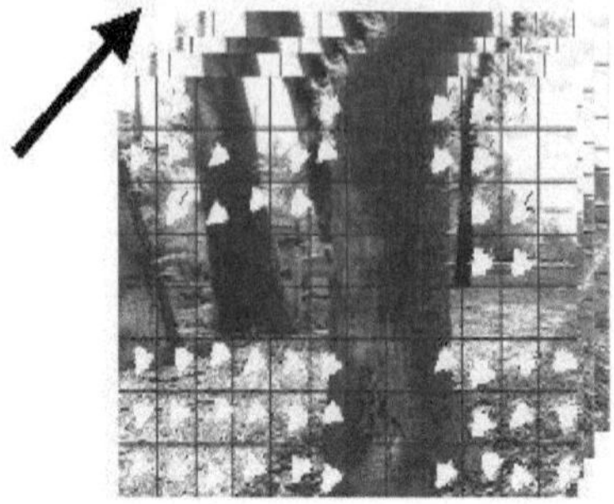

Fig. 2.2. Optical flow compressed and stored for a number of frames in the past

Fig. 2.3. (a) Points selected in initial video frame. (b) Points drawn in the locations from which they came 200 frames in the past.

a set of past camera images in which the vehicle was still some distance from its current location. Using the optical flow techniques discussed in this paper, a set of templates is assembled from the previous frames, each template representing the appearance of the most recent known road (the region immediately in front of the vehicle at the current time) at a different distance from the vehicle. Template matching in the current video frame performs road localization without requiring prior assumptions about the visual characteristics of the roadway. The process is repeated as the vehicle travels, which yields a self-supervised learning system which adapts to changing road conditions. The algorithm thus consists of three main parts: reverse optical flow, fast one-dimensional template matching, and dynamic programming. This process is shown in Fig. 2.5. The algorithmic steps needed for each application in addition to the optical flow calculations are discussed below.

2.2.1 Offroad Navigation

Image Segmentation: Using the short-range sensor suite with which the robot is equipped (infrared and physical bumpers) the robot can learn what objects are hazardous at close range either by bumping into them or coming very close to them. The obstacles and hazards can then be identified in the current frame based on *a priori* knowledge of where in the frame an obstacle which tripped

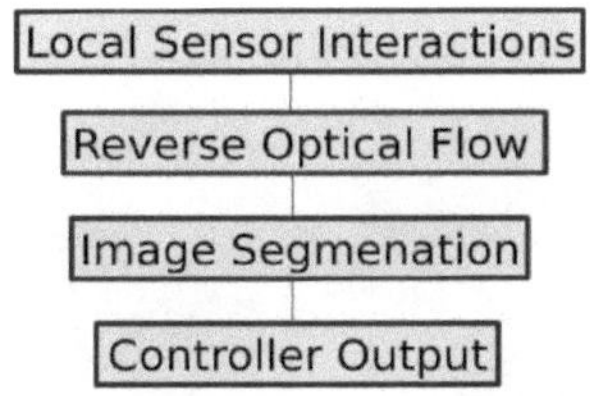

Fig. 2.4. Offroad navigation algorithm

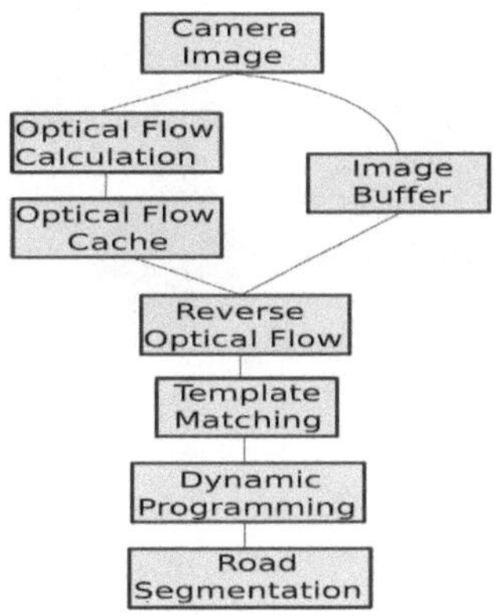

Fig. 2.5. Road following algorithm

the bumpers would be likely to lie. The reverse optical flow technique discussed previously is then employed to find the location of the current obstacle in a frame a set amount of time in the past.

Once a region in a previous frame has been selected that contains the local object at great distance, a matching measure is assembled using color and texture information from the object's distant appearance. Specifically, a color sample is obtained from a small window on the object, and a small image patch is stored as a texture sample from the same area. Once the color and texture samples have been taken from the object in the distant frame, each current image can be searched for regions exhibiting characteristics similar to the now known object. Each pixel of the current camera image is compared against the color sample taken from the object's distant appearance, and SSD (sum of squared differences) template matching is used to measure the similarity of the patch around each pixel with the stored texture from the object's distant appearance. In this way, once an object has been identified as a hazard at close range, it can be recognized and avoided at large distances.

2.2.2 Adaptive Road Following

Template Matching: This approach requires that the vehicle is traveling on the road to begin with and subsequently tracks image regions with visual

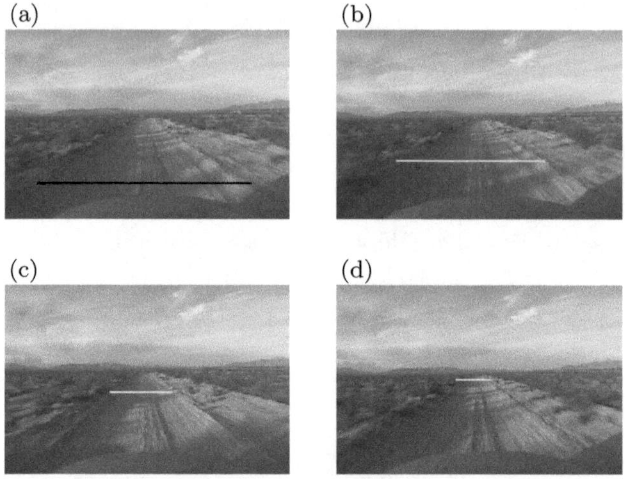

Fig. 2.6. (a) The location of the definition region in the current frame. (b)-(d) The locations in previous frames of the definition region (determined using reverse optical flow).

characteristics similar to the area directly in front of the vehicle, the *definition region*. The dark line in Fig. 2.6a shows a typical one-pixel-high definition region used in the algorithm. To identify areas of the current image which resemble the definition region via template matching, a set of horizontal templates are gathered that reflect the characteristics of the definition region at various distances in front of the vehicle. Because perspective and illumination effects alter the width, brightness, and texture of the definition region at different distances from the vehicle, an effective approach is to simply form the templates by pulling the image region corresponding to the current definition region from previous images when the region was more distant.

The reverse optical flow procedure described in this paper allows the location of the definition region in previous frames to be calculated. Taking the location of the traced-back definition region in a set of frames, each of which is progressively further in the past, provides a set of horizontal templates of the definition region at varying distances. Figs. 2.6(b-d) show the results of the reverse optical flow procedure applied to the definition region shown in Fig 2.6a. Horizontal templates such as those located along the white lines in Fig. 2.6(b-d) can then serve as cross-sectional templates used to determine the location of the road in the current image.

Using this set of horizontal templates that describe the visual characteristics of the definition region at various distances in front of the vehicle, a template matching algorithm can be used to search for the most likely position of the road at various vertical heights in the current frame. The vertical search heights are chosen as the same vertical heights from which the definition region templates were drawn in the past frames. However, changes in the terrain the robot is traveling over and its pitch can change the vertical height in the image of corresponding sections of roadway in different frames. To alleviate this effect, a Hough transform-based horizon detector is used to scale the vertical heights of the template search lines according to the vertical height of the horizon in the current image.

Since the templates and the search space are horizontal slices of the image, templates taken from curved roads appear very similar to those from straight road segments. Horizontal templates taken from roads with a different orientation than the vehicle's current path will be artificially wide, however, since the horizontal cross-section of the road is wider at these points. A similar effect would occur if the vehicle were undergoing modest amounts of roll. The template matching measure, combined with the dynamic programming procedure described below, mitigates the problems caused by these effects. If there is significant roll, however, errors in the horizon detector could adversely affect the accuracy of the algorithm.

An SSD measure is used to compute the goodness of the template match along each horizontal search line. The normalized SSD measure is defined as follows (where I is the image, T is the template, $\{x', y'\}$ range over the template, and $\{x, y\}$ range over the image):

$$R(x,y) = \frac{\sum_{x'} \sum_{y'} [T(x',y') - I(x+x', y+y')]^2}{[\sum_{x'} \sum_{y'} T(x',y')^2 \cdot \sum_{x'} \sum_{y'} I(x+x', y+y')^2]^{0.5}} \quad (2.1)$$

This matching measure can be calculated rapidly because the template is small (10-20 pixels) and the search space for each of the templates is a single horizontal line in the current image. An illustration of the matching response for a set of 10 templates is shown in Fig. 2.7a. The video frame itself is shown in Fig. 2.7b. The grayscale level represents the strength of the match, with white corresponding to a strong match, and black to a weak one. The responses in the figure have been widened vertically for clarity, but matching is only performed along a single horizontal line. It can be seen that strong responses occur in image regions near the center of the actual location of the road. Nevertheless, it can also be seen that strong responses sometimes also occur at other locations along each search line if the road is not clearly visually distinguishable from the rest of the scene, as is the case in the lower left portions of Fig. 2.7a and 2.7b, where the shadows and lack of vegetation combine to make some template matches to the left of the road attractive. This collection of SSD matching responses can also be used as a confidence measure. If the goodness of the best SSD measure drops sharply for all of the horizontal lines in the current image at once, it is likely that the vehicle has left the road or that the characteristics of the road have changed instantaneously. Actions could be taken at this point such as alerting a human operator or beginning an active search for areas near the vehicle with the attributes of the last known good road.

(a)

(b)

Fig. 2.7. (a) SSD response for 10 horizontal templates. Lighter colors correspon to stronger matches. (b) Corresponding input video frame.

Dynamic Programming: Fig. 2.8 shows the location of the maximum SSD response along each of the horizontal search lines with dark circles. While several of the circles correctly locate the actual position of the road, it is clear that the maximum response location is not necessarily correct in every case. Choosing the location of maximum response along each line would also allow estimates of the position of the road that might be physically unrealizable.

Finding the globally optimal set of estimated road positions while at the same time satisfying a constraint on the minimum radius of curvature of the road suggests the use of dynamic programming. Dynamic programming variants have been used already for aerial [14] and ground-based [15, 16, 17] road and lane detection. The goal of dynamic programming in this algorithm is to calculate the horizontal position of each template along the one-dimensional search lines so that when they are taken together the positions minimize a global cost function. The cost function used here is the arithmetic inverse of the SSD response at each horizontal position along the search lines, summed over all search lines. The search lines are processed from the topmost downward, with the cost at each horizontal position computed as the SSD cost at that location plus the minimum cost within a window around the current horizontal position in the search line above. The horizontal position of this minimum cost location is also stored as a link. The finite window restriction serves to enforce a constraint on the maximum allowable curvature of the road and also reduces the computation

Fig. 2.8. Black circles are the locations of maximum SSD response along the search lines. White circles are the output of the dynamic programming step. The dark line is the final output of the algorithm and is interpolated from the dynamic programming output.

time of the optimization. Once the bottommost search line has been processed in this way, the globally optimal solution is found by following the path of stored links, each of which point to the minimum cost position in the search line above. The path traversed represents the center of the estimated position of the road.

The output of the dynamic programming module is represented by the light circles in Fig. 2.8. The position of the entire road can now be estimated by interpolating between the optimal positions determined by the dynamic programming. The dark line in Fig. 2.8 indicates the estimated location of the center of the road. This was produced by interpolation using a 4th-degree polynomial fit of the dynamic programming output. The width of the segmented region was then linearly interpolated from the widths of the horizontal templates.

2.3 Results

The ability to recognize objects at a distance using computer vision after having identified them with a short-range sensor suite allows the use of self-supervised learning. The results of applying these optical flow techniques and self-supervised learning to offroad navigation for mobile robots and adaptive road following are discussed below.

2.3.1 Offroad Navigation

This application is part of a larger body of work being done on the DARPA Learning Applied to Ground Robots (LAGR) program [1]. The goal of the program is for the robot pictured in Fig. 2.10 to navigate autonomously offroad between two GPS waypoints while avoiding obstacles using computer vision as the only long-range sensor. Since the visual obstacles at the remote test sites where the races are run may vary drastically from obstacles used to test the vehicle on our test courses, self-supervised learning is an attractive approach.

This process is shown in Fig. 2.9. Fig. 2.9a shows a frame from a video sequence in which an obstacle (a tree) is close enough to be identified with the short-range sensor suite. Fig. 2.9b shows the location in a frame 200 frames in the past to which the selected point from Fig. 2.9a has been traced. Fig. 2.9c shows the results of simple color and texture matching using the appearance of the obstacle at long distance. White indicates a strong texture match while light gray indicates a color match. Trees tagged with color or texture matches in the current frame are labeled as obstacles and can be avoided using either an image space controller or a traditional planner after using a flat ground plane assumption to drop them into an occupancy grid. Fig. 2.9d shows the results of the simple color and texture matching using the appearance of the obstacle at the near distance when it was first recognized. The results are visibly worse, adding false positives around the base of the near tree and the park bench, and missing the far tree almost entirely. As expected, however, the matches for the two trees closest to the robot are better in Fig. 2.9d than in Fig. 2.9c while the matches for the three trees farther away are worse. It makes sense, therefore, to use a combination of templates for color and texture matching from several different time slices in the past to get the best possible matching results.

To test the effectiveness of this self-supervised learning approach we mounted a video camera on the iRobot ATRV robot shown in Fig. 2.11 and ran the algorithm described above. The result of the template matching step is then binarized to label everything in the scene as either hazardous or non-hazardous, as shown in Fig. 2.12a. Next a simple algorithm divides the binarized image into columns and determines which column extends the highest in the image before encountering a hazard. This column is then selected as shown by the vertical arrow in the sample frame in Fig. 2.12b. This information is passed to an image space controller that commands the robot to turn left or right based on the location of the arrow. Video of this simple system avoiding trees after a single training interaction with a tree can be found at *http://cs.stanford.edu/group/lagr/self-supervised.wmv.*

2.3.2 Adaptive Road Following

This section summarizes the results of previously published work done on an adaptive, self-supervised learning algorithm that targets ill-structured roads using a reverse optical flow technique and makes no prior assumptions about the visual characteristics of the roadway being followed [18].

The proposed algorithm discussed in Section II of this paper worked well in a variety of test conditions. Single frame results from three different 720 x 480 pixel video sequences taken in the Mojave Desert are shown in Fig. 2.13. Each column of Fig. 2.13 contains results from a different one of the data sets. The first video sequence contains footage of a straight dirt road in a sparse desert environment and illustrates the tendency of the algorithm to correctly segment the road even in environments containing many regions with similar visual characteristics to

Fig. 2.9. (a) Appearance of tree within range of short-range sensors. (b) Location of tree 200 frames in the past. (c) Result of color and texture matching based on appearance of tree at far distance, in current frame. (d) Result of color and texture matching based on appearance of tree at near distance, in current frame.

Fig. 2.10. The LAGR robot platform

the road surface. The second sequence is video of a gravel road winding up and down a rocky hill in bright sunshine. The third sequence has footage of a curved road late in the afternoon when shadows began to fall on the roadway.

The following are the details of the implementation used to produce the output images and videos discussed in this paper. The road position estimates are the result of 1-D template matching of a set of 10 horizontal templates acquired using the reverse optical flow procedure. The templates are samples at different points in time of a definition region directly in front of the vehicle. They were taken from times in the past ranging from 1 frame to roughly 200 frames prior to the current

Fig. 2.11. iRobot ATRV platform

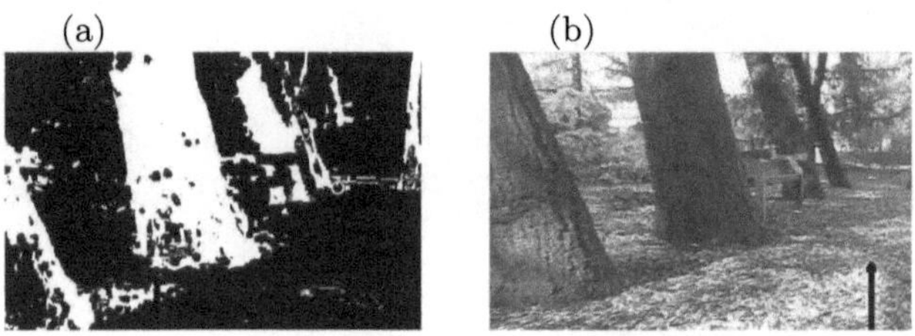

Fig. 2.12. (a) Binarized hazard image. (b) Arrow to indicate action controller will take.

frame. The particular temporal samples were chosen to provide an evenly spaced set of horizontal templates in the image plane. Each template is 20 pixels high, and the definition region and templates were refreshed every 10 frames to accommodate gradual changes in road appearance. Optical flow fields were measured using 3000 feature correspondences and mean flow vectors were stored in a grid of 96 rectangular cells covering the entire camera image. Interpolation of the dynamic programming output was achieved using a 4th-degree polynomial fit.

To investigate the performance of the algorithm on real-world data, we evaluated the results of the three 1000-frame data sequences discussed above using two performance metrics described here.

Pixel Coverage Metric: The first metric compares pixel overlap between the algorithm output and ground truth images in which the road has been segmented by a human operator, as shown in Fig. 2.14. The number of pixels in the frame that have been incorrectly labeled as roadway is subtracted from the number of correctly labeled roadway pixels. This number is then divided by the total number of pixels labeled as road by the human operator for that frame. Using the metric proposed here, a score of 1.0 would correspond to correctly identifying all the road pixels as lying in the roadway, while not labeling any pixels outside

Fig. 2.13. Sample single frame algorithm outputs for the three data sets. Each column contains results from one of the three video sequences.

the roadway as road pixels. A score of 0.0 would occur when the number of actual road pixels labeled as roadway is equal to the number of non-roadway pixels incorrectly identified as being in the road. If more pixels were incorrectly labeled as roadway than actual road pixels correctly identified, negative scores would result. This measure is computed once per frame and averaged over the entire video sequence. While this pixel coverage metric is easily visualized and simple to compute, it must be recognized that, due to perspective effects, it is strongly weighted towards regions close to the vehicle.

Fig. 2.14. Sample human-labeled image from test sequence used for test metrics

Line Coverage Metric: The second metric mitigates the distance-related bias of the first metric by comparing pixel overlap separately along a set of horizontal lines in the images. Five evenly spaced horizontal lines are chosen ranging in vertical position between the road vanishing point and the vehicle hood in the ground-truth image. Success scores are calculated just as in the first metric, except they are reported individually for each of the five lines. The metric returns five sets of success scores computed once per frame and averaged over the entire video sequence.

The scores for the algorithm on each of the three video sequences, evaluated using the pixel coverage metric were as follows. 0.6958 for the straight road video, 0.6983 for the video with the curved road climbing the hill, and 0.6322 for the video of the curved road with shadows.

The strengths of the proposed algorithm are best highlighted in the first test video sequence, which consists mainly of a straight dirt road surrounded by large non-road regions with visual characteristics strikingly similar to that of the road. Despite the fact that the road is only loosely defined with respect to the surrounding regions, our algorithm is able to correctly locate the roadway with a strong pixel coverage metric score.

The second test video sequence, which contains a gravel road curving up a hill, presents no major problem for our algorithm. Curved roads with significant elevation changes do not seem to adversely affect our algorithm, as compared to the straight road found in the first test sequence.

The addition of intermittent shadows as found in the third test sequence does slightly affect the performance of our algorithm. This effect is the result of the fact that a template taken from a region in an image in shadow will not produce strong matches when moved across a line in the current image which is not in shadow.

Fig. 2.15 shows the performance of the algorithm on the same three data sets, now evaluated using the line coverage metric. Metric scores are graphed for a set of five evaluation lines increasingly distant from the vehicle. As could be expected, the performance of the algorithm generally declines as the distance from the vehicle increases. The proposed algorithm achieves very low false positive rates by making no assumptions about the general appearance of the road and following regions that adhere to its learned roadway information. The inability of our algorithm to achieve high rates of correct roadway labeling near the horizon is at least in part due to the fact that optical flow records are only stored for a fixed number of frames in the past. Therefore the definition region can never be traced back all the way to the horizon.

Videos of the three 1000-frame test sets, showing the results of tracking with the proposed algorithm as well as with our implementations of naive color and texture algorithms for our own comparison purposes, are available at *http://cs.stanford.edu/group/lagr/road_following/*. The algorithm runs at 3Hz on a 3.2GHz PC at 720 x 480 pixel resolution. For a more detailed description of the test results as well as a discussion of limitations of the algorithm and future work,please reference the original paper [18].

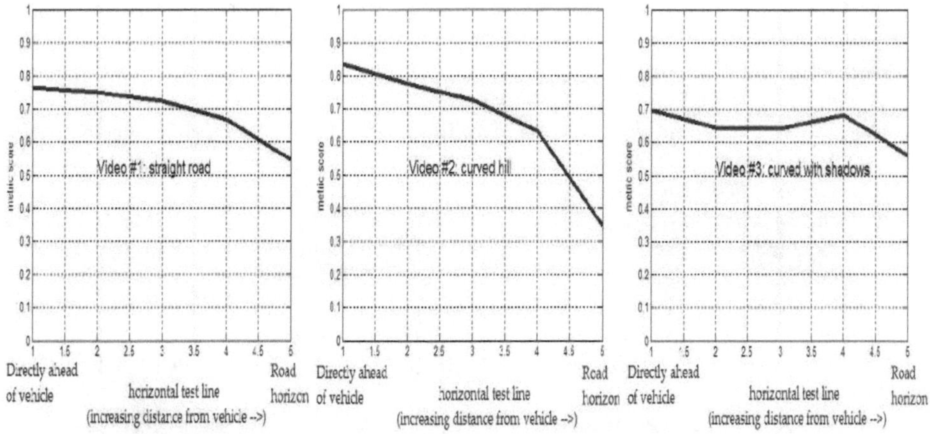

Fig. 2.15. Pixel coverage results are shown as a function as distance from the vehicle in image space for the three video sequences

2.4 Conclusion

This paper covers work done in our lab on optical flow techniques which allow mobile robot navigation via self-supervised learning. By using optical flow to correlate the appearance of obstacles when they trigger the local sensors of a robot with their appearance at some distance from the robot, this approach enables efficient obstacle avoidance. Two applications of this work were discussed: autonomous off-road navigation and adaptive road following.

Acknowledgments

This research has been financially supported through the DARPA LAGR program. The views and conclusions contained in this document are those of the authors, and should not be interpreted as necessarily representing policies or endorsements of the US Government or any of the sponsoring agencies. The authors would like to acknowledge the Stanford Racing Team for providing the test videos used in the preparation of this paper.

References

[1] Defense Advanced Research Projects Agency (DARPA), Learning Applied to Ground Robots (LAGR), Online source: http://www.darpa.mil/ipto/programs/lagr/vision.htm.

[2] Murray, D. and Little, J., "Using Real-Time Stereo Vision for Mobile Robot Navigation," *Autonomous Robots*, vol. 8, Issue 2, pp. 161-171, Apr. 2000.

[3] DeSouza, G. and Kak, A., "Vision for Mobile Robot Navigation: A Survey," *IEEE Transactions on Pattern Analysis and Machine Intelligence*, vol. 24, no. 2, pp. 237-267, Feb. 2002.

[4] Moorehead, S., Simmons, R., Apostolopoulos, D., and Whittaker, W. L., "Autonomous Navigation Field Results of a Planetary Analog Robot in Antarctica," *International Symposium on Artificial Intelligence, Robotics and Automation in Space*, June 1999.
[5] Asensio, J. R., Montiel, J. M. M., Montano, L., "Goal Directed Reactive Robot Navigation with Relocation Using Laser and Vision," *IEEE Proc. of International Conference on Robotics and Automation*, vol 4, pp. 2905-2910, 1999.
[6] Ulrich, I. and Nourbakhsh, I., "Appearance-Based Obstacle Detection with Monocular Color Vision," *Proceedings of AAAI Conference*, pp. 866-871, 2000.
[7] Lorigo, L. M., Brooks, R. A., Grimson, W. E. L., "Visually-Guided Obstacle Avoidance in Unstructured Environments," *In Proceedings of the IEEE/RSJ International Conference on Intelligent Robots and Systems*, pp. 373-379, 1997.
[8] Horswill, I., "Polly: A Vision-based Artificial Agent," *Proceedings of AAAI Conference*, pp. 824-829, 1993.
[9] Chirkhodaie, A. and Amrani, R., "Visual Terrain Mapping for Traversable Path Planning of Mobile Robots," *Proceedings of SPIE*, vol. 5608, pp. 118-127, Oct. 2004.
[10] Coombs, D., Herman, M., Hong, T., and Nashman, M., "Real-Time Obstacle Avoidance Using Central Flow Divergence, and Peripheral Flow," *IEEE Trans. on Robotics and Automation*, vol. 14, no. 1, pp. 49-59, 1998.
[11] Giachetti, A., Campani, M., and Torre, V., "The Use of Optical Flow for Road Navigation," *IEEE Trans. on Robotics and Automation*, vol. 14, no. 1, pp. 34-48, 1998.
[12] Shi, J. and Tomasi, C., "Good Features to Track," *Proc. of the IEEE Conference on Computer Vision and Pattern Recognition*, pp. 593-600, 1994.
[13] Bouguet, J., "Pyramidal Implementation of the Lucas Kanade Feature Tracker Description of the Algorithm" Intel Corporation, Microprocessor Research Labs, 2000. OpenCV Documents.
[14] Dal Poz, A. P., do Vale, G. M., "Dynamic programming approach for semi-automated road extraction from medium- and high-resolution images," *ISPRS Archives*, vol. XXXIV, part 3/W8, Sept. 2003.
[15] Redmill, K., Upadhya, S., Krishnamurthy, A., Ozguner, U., "A Lane Tracking System for Intelligent Vehicle Applications," *Proc. IEEE Intelligent Transportation Systems Conference*, 2001.
[16] Kim, H., Hong, S., Oh, T., Lee, J., "High Speed Road Boundary Detection with CNN-Based Dynamic Programming," *Advances in Multimedia Information Processing - PCM 2002: Third IEEE Pacific Rim Conference on Multimedia*, pp. 806-813, Dec. 2002.
[17] Kang, D., Jung, M., "Road Lane Segmentation using Dynamic Programming for Active Safety Vehicles,"*Elsevier Pattern Recognition Letters* vol. 24, issue 16, pp. 3177-3185, 2003.
[18] Lieb, D., Lookingbill, A., and Thrun, S.,"Adaptive Road Following using Self-Supervised Learning and Reverse Optical Flow," *Proceedings of Robotics: Science and Systems*, 2005.

3

Steps Towards Safe Navigation in Open and Dynamic Environments

Christian Laugier, Stéphane Petti, Dizan Vasquez, Manuel Yguel, Thierry Fraichard, and Olivier Aycard

INRIA Rhône-Alpes & GRAVIR Lab. (CNRS, INPG, UJF)
http://emotion.inrialpes.fr

Summary. Autonomous navigation in open and dynamic environments is an important challenge, requiring to solve several difficult research problems located on the cutting edge of the state of the art. Bassically, these problems can be classified into three main categories: SLAM in dynamic environments; Detection, characterization, and behavior prediction of the potential moving obstacles; On-line motion planning and safe navigation decision based on world state predictions. This paper addresses some aspects of these problems and presents our latest approaches and results. The solutions we have implemented are mainly based on the followings paradigms: *Characterization and motion prediction* of the observed moving entities using bayesian programming; *On-line goal-oriented navigation decisions* using the Partial Motion Planning (PMP) paradigm.

3.1 Introduction

3.1.1 Outline of the Problem

To some extent, autonomous navigation for robotic systems placed in stationary environments is no longer a problem. The challenge now is autonomous navigation in open and dynamic environments, *i.e.* environments containing moving objects (potential obstacles) whose future behaviour is unknown. Taking into account these characteristics requires to solve several difficult research problems at the cutting edge of the state of the art. Basically, these problems can be classified into three main categories:

- Simultaneous Localisation and Mapping (SLAM) in dynamic environments;
- Detection, tracking, identification and future behaviour prediction of the moving obstacles;
- On-line motion planning and safe navigation

In such a framework, the system has to continuously characterize the fixed and moving objects that can be observed both with on-board or off-board sensors. As far as the moving objects are concerned, the system has to deal with problems such as interpreting appearances, disappearances, and temporary occlusions of rapidly manoeuvring objects. It also has to reason about their future behaviour (and consequently to make predictions). From the autonomous navigation point

C. Laugier and R. Chatila (Eds.): Autonomous Navi. in Dyn. Environ., STAR 35, pp. 45–82, 2007.
springerlink.com

of view, this means that the system has to face a double constraint: constraint on the response time available to compute a safe motion (which is clearly a function of the dynamicity of the environment), and a constraint on the temporal validity of the motion planned (which is a function of the validity duration of the predictions). In other words, one needs to be able to plan motion fast, but one does not need to plan motion very far in the future.

This paper addresses some aspects of the previous problem, and presents our latest approaches and results. The solutions we have implemented rely on the following modules:

- *Scene interpretation and short-term motion prediction* for the moving obstacles, using the new concept of Bayesian Occupancy Filters and an efficient wavelet-based representation of the related occupancy grids;
- *Medium-term motion and behaviour prediction* for the observed moving objects, using motion pattern learning and Hidden Markov Models;
- *On-line goal-oriented navigation decision* using the Partial Motion Planning (PMP) paradigm.

3.1.2 Case Study: The Automated Valet Parking

One possible (and very relevant) target application for the techniques presented in this paper is that of the Automated Valet Parking (AVP). The robotic system considered is a "smart" car operating autonomously in a "smart" city parking. Both the car and the parking are equipped with sensors providing them with information about the world. Let us imagine the following scenario: you drive your car and leave it at the entrance of a given parking. From then on, it operates autonomously and go park itself. As soon as the car enters the parking, the car on-board intelligent system connects to the parking's own intelligent system and request a free parking place. The parking then confirms the availability of a parking space and provides the car with a model of the parking and an itinerary to the empty place. From then on, the car, using information obtained from both its own sensors and the parking sensory equipment, go park itself.

From an architecture point of view, the AVP scenario involves two "intelligent" systems communicating with one another: the car on-board system and the parking off-board system. As mentioned earlier, it is assumed that both systems are equipped with sensors providing them with information about the environment considered. While the car sensors will provide it with a local view of its surroundings, it can be expected that the parking sensors will provide the car with an overall view of what is going on in the whole parking.

To address the AVP scenario, we have devised a scheme relying upon the following functionalities (split between the car and the parking).

Parking abilities

- Parking monitoring: at any time, the parking should know which places are free or not (and by whose car they are occupied).

- Route planning: the parking should be able to provide the car with a model of the parking premises along with the best itinerary to reach a given place.
- Moving objects monitoring: any moving objects (vehicles, pedestrians, etc.) should be monitored and tracked. The parking should be able to provide information such as position, speed and expected trajectory (*i.e.* future behaviour). Expected trajectories can come from different clues: typical movements of the different kinds of moving objects, learnt from previous observation, or knowledge of a planned trajectory.
- Car localisation: given its moving objects' monitoring functionality, the parking can provide the car with its current state in the parking premises.

Car abilities

- Localisation: the car should be able to maintain an estimate of its localisation in the parking. It can be the result of a data fusion between parking information and on-board localisation.
- Environment modelling: the car on-board sensor are primarily used to build a model of the surroundings of the car. This local model should be enriched using the global information provided by the parking (in particular, the information concerning the moving objects' future behaviour).
- Automated driving: given the parking model and the route to the goal, the car should be able to determine its future course of action so as to reach its goal efficiently and safely.

One can notice that some of the functionalities mentioned above are somewhat redundant (in particular when dealing with sensing data). This property has intentionally been chosen in order to increase the robustness and the efficiency of the system:

- Fusion of data from multiple source increase overall accuracy.
- Using several data source increase fault tolerance.
- By correlating different inputs, it is possible to diagnose if an input is failing or becoming unreliable.

3.1.3 Outline of the Paper

In this paper, we focus on two of the functionalities mentioned in the previous section: *Motion prediction of the observed moving objects* and *on-line goal-oriented navigation decisions*. The paper is organized in three main sections. The section 3.2 describes how we have solved the problem of interpreting and representing the dynamic environment of the robot using the "Bayesian Occupancy Filtering"(*BOF*) approach; this approach relies on a local world-state bayesian interpretation scheme, including a short-term motion prediction mechanism. The section 3.3 deals with the problem of the prediction of the most likely behaviors of some observed objects executing "intentional motions"; the proposed solution relies on the use of a motion pattern learning mechanism and of an extended Hidden Markov Model. The section 3.4 deals with the problem of planning safe motions in a reconstructed dynamic environment; the proposed paradigm (called "Partial Motion Planning", or PMP) takes into account (at each iteration step) both the time constraints and the current model of the future state of the robot environment.

3.2 Scene Interpretation and Short-Term Motion Prediction

3.2.1 Overview of the Problem

The problem addressed in this section concerns the interpretation of the observed dynamic scene in terms of *potential moving obstacles*, i.e. obstacles which may generate a collision in the near future with the robot). The objective is to be able to correctly interpret the dynamic scene in the presence of noisy or missing data, and to be as robust as possible to temporary or partial occlusions. Our approach for solving this problem is based on the new concept of *Bayesian Occupancy Filtering (BOF)* [1], where the robot environment is represented using a *4-dimensional occupancy grid*, i.e. an occupancy grid which includes the velocity dimension.

The *occupancy grids* [2, 3] framework is a classical way to describe the environment of a mobile robot. It has extensively been used for static indoor mapping [4] using a 2-dimensional grid. More recently, occupancy grids have been adapted to track multiple moving objects [5]. However, a major drawback of these approaches, is that a moving object may be lost due to occlusion effects.

The BOF approach avoid this problem for short temporary occlusions (e.g. a few seconds), by combining two complementary phases in a recursive loop: the *estimation phase* which estimate the occupancy probability of each cell of the 4-dimensional grid, using recursively the set of sensor observations; the *prediction phase* which estimate an a priori model of the grid occupancy at time $k+1$, using a "dynamic model" and the latest estimation of the grid state (figure 3.2 illustrates). This approach has been developed using the *Bayesian Programming* Framework [6, 7, 8]; it is described in the next sections.

However, large scale environments can hardly been processed in real-time because of the intrinsic complexity of the related inferences and numerical computations (see section 3.2.6). The section 3.2.7 presents the outline of the WOG model ("Wavelet Occupancy Grid") we are developing for trying to meet the required efficiency property.

3.2.2 Estimation of the Occupancy Probability

The *estimation phase* consists in estimating, at each time step, the occupancy probability of each cell of the 4-dimensional grid. This estimation is performed using recursively the set of "observations" (i.e. pre-processed sensors data) provided by the sensors at each time step. These observations are represented by a list of detected objects, along with their associated positions and velocities in the reference frame of the processed sensor (several sensors may be used in parallel). In practice, this set of observations could also contain two types of false measurements : the *false alarms*, i.e. when the sensor detects a non existing object; the *missed detection*, i.e. when the sensor does not detect an existing object.

Solving the related estimation problem is done by using the available instantaneous information about the environment state (i.e. the current observations and grid state). A sketch of the algorithm is given below using our *Bayesian Programming* Framework [6, 7, 8]; a more complete description of the method (which includes a "sensor data fusion" step) can be found in [1].

(i) Choosing the relevant variables and decomposition

- O_{x,y,v_x,v_y}: The occupancy of the cell (x, y, v_x, v_y) at time t: occupied or not. This variable is indexed by a 4-dimensional index that represents a position and a speed relative to the vehicle.
- $\mathcal{Z}$: The sensor observation set; one observation is denoted Z_s; the number of observation is denoted S.

If we make the reasonable assumption that all the observations of the sensors are independent when knowing the occupancy state of a cell, we can choose to apply the following decomposition of the related joint distribution:

$$P(O_{x,y,v_x,v_y}, \mathcal{Z}) = P(O_{x,y,v_x,v_y}) \times \prod_{s=1}^{S} P(Z_s \,|O_{x,y,v_x,v_y}). \tag{3.1}$$

(ii) Assigning the parametric forms
According to our knowledge of the problem to be solved, we can assign the following parametric forms to each of the terms of the previous decomposition:

- $P(O_{x,y,v_x,v_y})$ represents the *a priori* information on the occupancy of each cell. If available, a *prior* distribution could be used to specify it. Otherwise, a uniform distribution has to be selected. The next section will show how the prior distribution may be obtained from previous estimations.
- The shape of $P(Z_s \,|O_{x,y,v_x,v_y})$ is given by the *sensor model.* Its goal is to model the sensor response knowing the cell state. Details about this model can be found in [3]; an example is given for a telemetric sensor in Fig. 3.1. Such models can easily been built for any kind of sensor.

(iii) Solution of the problem
It is now possible to ask the *Bayesian question* corresponding to the searched solution (i.e. the searched probability distribution). Since the problem to solve consists in finding a good estimate of the cell occupancy, the question can be stated as follows:

$$P(O_{x,y,v_x,v_y} \mid \mathcal{Z})? \tag{3.2}$$

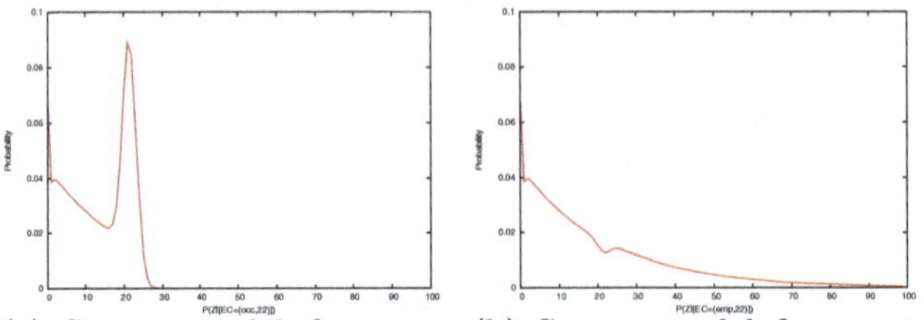

(a) Sensor model for occupied cells (b) Sensor model for empty cells

Fig. 3.1. Example of one-dimensional sensor models. The sensor is a laser-range finder located in $x = 0$ and detecting an object at $x = 22$. The following property holds : $P(Z|[O_x = occ]) = P(Z|[O_x = emp])$ for $x > z$, which implies that $P(O_x|Z)$ is unknown when the cell is behind the obstacle.

The result of the related Bayesian inference[1] can be written as follows:

$$P(O_{x,y,v_x,v_y} \mid \mathcal{Z} \,) \propto \prod_{s=1}^{S} P(Z_s \mid O_{x,y,v_x,v_y} \,). \tag{3.3}$$

3.2.3 The Bayesian Occupancy Filter

We are now interested in taking into account the sensor observation history, in order to be able to make more robust estimations in changing environments (i.e. in order to be able to process temporary objects occlusions and detection problems). A classical way to solve this problem is to make use of Bayes filters [9]. Basically, the goal of such a filtering process is to recursively estimate the probability distribution $P(O^k_{x,y,v_x,v_y} \mid Z^k)$, known as the *posterior* distribution. In general, this estimation is done in two stages: a *prediction* stage whose purpose is to compute an *a priori* estimate of the target's state known as the *prior* distribution; an *estimation* stage whose purpose is to compute the *posterior* distribution, using this *a priori* estimate and the current measurement of the sensor. Exact solutions to this recursive propagation of the posterior density do exist in a restrictive set of cases (such as the Kalman filter [10][11] when linear functions and gaussian distributions are used).

Our approach for solving this problem, is based on the new concept of *Bayesian Occupancy Filter (BOF)*; Fig. 3.2 shows the related estimation loop. This approach is derived from the general bayesian filtering paradigm; it provides a powerfull formalism for solving difficult estimation problems expressed in our 4-dimensional occupancy grid representation.

The basic idea underlying the conception of the *BOF* is to make use of the velocities measured in the past, for predicting the near future and propagating this information through time. Indeed, knowing the current velocity of a mobile,

[1] This result is computed using our *ProBT* inference engine, currently commercialized by our spin-off company *Probayes*.

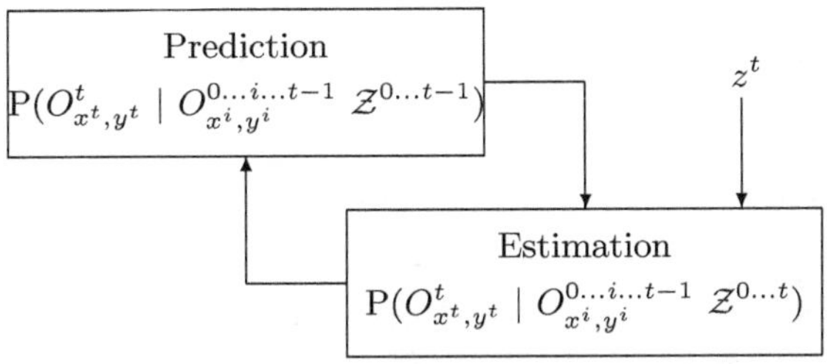

Fig. 3.2. Bayesian Occupancy Filter as a recursive loop

it is possible to predict its future motion under the hypothesis of a constant velocity for the next time step (in practice, possible velocity changes will generate several possible future positions).

A complete presentation of the *BOF* can be found in [1]. In the sequel, we will just give an overview of the approach under the following simplifying assumptions (for clarity reasons) : use of a single sensor and constant velocity for the observed moving objects. A consequence of this last assumption, is that we can deal with a single "antecedent cell" when evaluating the occupancy state of a given cell.

(i) Choosing the relevant variables and decomposition

- $O^t_{x,y}$: Occupancy of the cell $c = (x, y, v_x, v_y)$ at time t, occupied or not.
- $O^{t-\delta t}_{x_p,y_p}$: Occupancy of the cell which is the antecedent of $C^t_{x,y}$, occupied or not. In this model, $x_p = x - v_x\delta t$ and $y_p = y - vy\delta t$, since the velocity is constant.
- Z_s : A sensor observation.

Under the previous simplifying assumptions, the following decomposition of the joint distribution determined by these variables can be expressed as follow:

$$P(O^{t-\delta t}_{x_p,y_p}, O^t_{x,y}, Z_s) = \left(\begin{array}{l} P(O^{t-\delta t}_{x_p,y_p}) \\ \times P(O^t_{x,y}|O^{t-\delta t}_{x_p,y_p})P(Z_s|O^t_{x,y}) \end{array} \right). \tag{3.4}$$

(ii) Assigning the parametric forms

- $P(O^{t-\delta t}_{x_p,y_p})$ is the *prior* for the future occupancy state of the cell $c = (x, y, v_x, v_y)$. For each cell c such as the antecedent (x_p, y_p, v_x, v_y) is out of the current grid, this prior is the probability that a new object enters in the monitored space; since we usually have no real information about such an event, this probability is represented by a uniform distribution.
- $P(O^t_{x,y}|O^{t-\delta t}_{x_p,y_p})$ is related to the very simple "dynamic model" we are using in this case. It is defined as a transition matrix $\begin{bmatrix} 1-\epsilon & \epsilon \\ \epsilon & 1-\epsilon \end{bmatrix}$, which allows the

system to take in account the fact that the null acceleration hypothesis is an approximation; in this matrix, ϵ is a parameter representing the probability that the object in $c = (x_p, y_p, v_x, v_y)$ does not follow the null acceleration model.

- $P(Z_s|O^t_{x,y})$ is the sensor model (see section 3.2.2).

(iii) Solution of the problem.
Similarly to the estimation process described in the section 3.2.2, the solution of the problem to be solved by the BOF can be defined by the following Bayesian question : $P(O^t_{x,y}|Z_s)$?. Answering this question (i.e. computing the related probability distribution) is achieved using our inference engine; this inference involves a marginalization sum over $O^{t-\delta t}_{x_p,y_p}$.

3.2.4 Experimental Results

This approach has been implemented and tested on our experimental platform : the Cycab vehicle equiped with a laser sensor. Fig. 3.3 shows some resulting grid estimations, for an environment containing two stationary objects and an object moving from the left to the right at a velocity of 0.8 m/s; in this example, the robot is not moving.

Fig. 3.3b depicts the occupancy probability of each cell corresponding to a null relative velocity (*i.e.* $c = [x, y, 0, 0]$). As expected, two areas with high occupancy probabilities are visible; these probability values depends on several factors attached to the sensor model : the probability of true detections, the probability of false alarms, and the sensor accuracy. Since the measured speed for the third obstacle is not null, any area of high occupancy probability corresponding to this observation is only represented in the related slices of the grid (i.e. the slice corresponding to $c = [x, y, 0, 0.8]$ in this case, see Fig. 3.3c).

It should be noticed that the cells located outside of the sensor field of view, or the cells "hidden" by one of the three sensor observations (i.e. the cells located behind the three detected obstacles) cannot be observed; consequently, nothing really consistent can be said about these cells, and the system has given an occupancy probability value of 0.5 for these cells.

Fig. 3.4 shows a sequence of successive prediction and estimation results given by the BOF. The experimental scenario involves a stationary obstacle and the Cycab moving forward at a velocity of 2.0 m/s. The obstacle is detected using the laser sensor; it finally goes out of the sensor field of view (see Fig. 3.4-d1), since the Cycab is moving forward. It should be noticed that the prediction step allows to infer knowledge about the current occupancy state of the cycab environment, even if the object is no longer observed by the sensor; this is the situation depicted by fig 3.4-d3, where an area of high occupancy probability still exists when the object is going out of the sensor field of view. In some sense, our prediction step can be seen as a "short-term memory", which allows to combine in an evolutive way past and current observations.

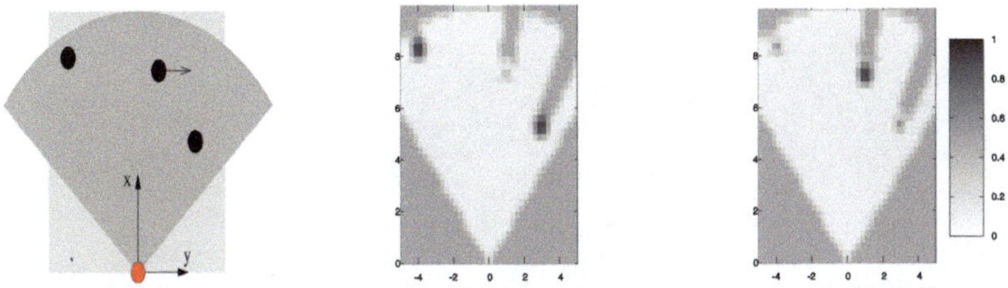

a) $z_1 = (8.3, -4, 0, 0)$ $z_2 = (5, 3, 0, 0)$ $z_3 = (7.3, 1.9, 0, 0.8)$

b) $P([O_C = occ] \mid \mathcal{Z})$ $[c = (x, y, 0, 0)]$

c) $P([O_C = occ] \mid \mathcal{Z})$ $[c = (x, y, 0, 0.8)]$

Fig. 3.3. Example of a grid estimation using a single laser sensor located at $(0, 0)$, in a scene including two stationary objects and an object moving from the left to the right at a velocity of 0.8 m/s. In this example, the robot is not moving. The occupancy probability value of each cell is represented by a gray level (see the colors correspondences on the right side of the figure). (a) The sensed environment and the three instantaneous sensor observations expressed in the $(x, y, \dot{x}, \dot{y})$ space. (b) Occupancy probability in a 2-dimensional slice of the grid corresponding to a null relative velocity (*i.e.* $c = [x, y, 0, 0]$). (c) Occupancy probability in a 2-dimensional slice of the grid corresponding to a relative horizontal velocity 0.8 (*i.e.* $c = [x, y, 0, 0.8]$).

3.2.5 *BOF* Based Collision Avoidance

In [1], we have shown how to avoid a mobile obstacle by combining the occupancy grid result given by the *BOF*, with a *danger probability* term computed using a "time to collision" criteria. In this approach, each cell O^k_{x,y,v_x,v_y} of the grid is characterized by two probability distributions : the "occupancy" term $P(O^k_{x,y,v_x,v_y} \mid \mathcal{Z}^k)$ and the "danger" term $P(D^k_{x,y,v_x,v_y} \mid O^k_{x,y,v_x,v_y})$. Using this model, it becomes possible to tune the velocity controls of the Cycab, according to the results of a combination of the two previous criteria.

This approach has experimentally been validated using the following scenario : the Cycab is moving forward; a pedestrian is moving from right to left, and during a small period of time the pedestrian is temporarily hidden by a parked car. Fig 3.5 shows some snapshots of the experiment : the Cycab brakes to avoid the pedestrian which is temporarily hidden by the parked car, then it accelerates as soon as the pedestrian has crossed the road.

3.2.6 Discussion and Performances

Thanks to the previous assumptions, both the prediction step and the estimation step complexities *increases linearly with the number of cells* of the grid. This make the approach tractable in real situations involving reasonable grid sizes. This is the case for the experimental scenarios described in this section (e.g. size of the sensed space of 10 m m with a resolution of 0.5 m, longitudinal

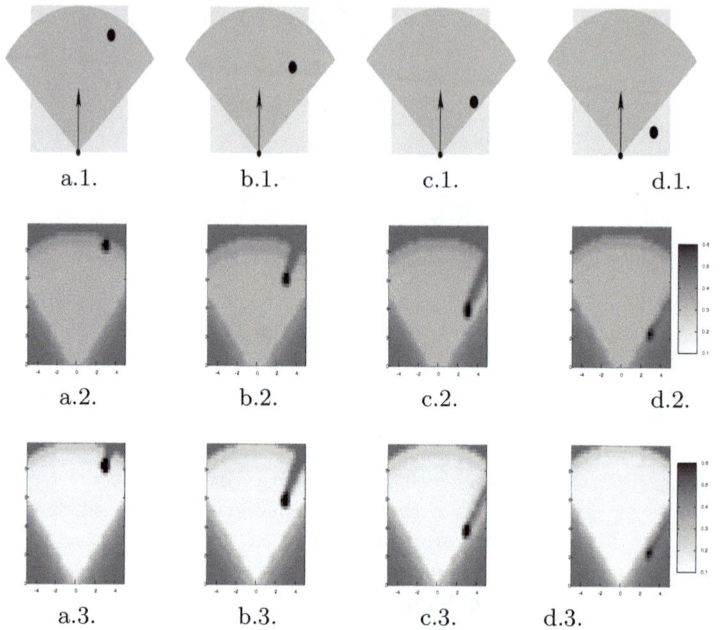

Fig. 3.4. A short sequence of a dynamic scene involving a stationary obstacle and the Cycab moving forward at a velocity of 2.0 m/s. The second and the third row respectively show the results of the prediction and of the estimation steps of the BOF, in a 2-dimensional slice of the grid corresponding to a relative speed of $\dot{x} = -2.0$, i.e. $P([O^t_{x\ y\ [\dot{x}=-2.0]\ [\dot{y}=0.0]} = occ]\ )$.

Fig. 3.5. Snapshots of the experimental pedestrian avoidance scenario

velocity of $-3\ m.s^{-1}$ to $1\ m.s^{-1}$ with a resolution of $0.4\ m.s^{-1}$, lateral velocity of $-3.2\ m.s^{-1}$ to $3.2\ m.s^{-1}$ with a resolution of $0.4\ m.s^{-1}$). Using such a grid of 64.000 cells, the computation time for both prediction and estimation steps is about 100 ms on a 1 GHz computer. This is fast enough to control the CyCab at a maximum speed of $2\ m.s^{-1}$.

However, this grid size is not fine enough for some other large scale applications involving higher speeds. In this case, the number of cells increases quickly, and the required computational time becomes too high for satisfying the real-time constraint.

In order to try to solve this problem, we are working in two directions. The first one consists in developing a dedicated hardware exploiting the *highly parallel* structure of the *BOF* algorithm[2]. The second approach consists in using a multi-resolution framework for representing the 4-dimensional grid and the related processing models. The outline of this approach is described in the next section.

3.2.7 Wavelet-Based Model for the *BOF* (the *WOG* Model)

(i) Overview of the problem

The goal of this new model is to provide a "coarse-to-fine" representation of the underlying *BOF* model, in order to be able to optimize the processing of large homogeneous regions, and to refine the model only when this is necessary. We have chosen to make use of the wavelet framework [12] [13], which allows the processing of large sets of data including non-linearities (as it is the case for our dynamic maps). Pai and Reissell [14] have shown that wavelets could be used for representing 3D static maps for path planning. Sinopoli [15] has extended this approach for solving a global path-planning problem, while using traditional 3D Occupancy Grids (*OG*) for local navigation.

Our approach, called *"Wavelet Occupancy Grid"* (*WOG*), can be seen as a tight combination of wavelet and *OG* representations, allowing us to perform *OG updates in the wavelet space*, and later on to make "prediction inferences" within the wavelet space (i.e. to fully implement the *BOF* paradigm in this multi-resolution formalism).

(ii) The WOG model

The first objective is to develop a wavelet-based model for 2-dimensional *OG* (i.e. without representing velocities). At this step of the development, we will only consider the random variables $O^t_{x,y}$ and Z^t (defined above). Since each occupancy variable ($O_{x,y}$) is binary, its probability distribution is completely defined by $P([O^t_{x,y} = occ])$. Then, we consider now the occupancy function $p^t(x, y) = P([O^t_{x,y} = occ])$ which allows to capture the space homogeneity.

The used wavelet model

Basically, linking *OG* and wavelet representations, leads to project a huge function representing the *OG* into a wavelet vector space. In our approach, we have used the 2D Haar model [13], built upon two types of basis functions: the "scale" and "detail" functions, where the scaling mother function is defined as follows:

$$\Phi(x, y) = 1 \text{ for } (x, y) \in [0, 1]^2, \text{ zero elsewhere}$$

[2] Thanks to the hypothesis that each cell is independent, the state of each cell can be computed independently.

Then, the Haar basis at scale s is defined as the union of the set of "scaled" functions $\{\Phi^{sij}|(i,j) \in \mathbb{Z}^2\}$ and the set of "details" functions $\{\Psi_M^{lij}|l = 0,\ldots,s;(i,j) \in \mathbb{Z}^2\}$, where:

$$\Phi^{lij} = 2^{-l}\Phi(2^{-l}x - i, 2^{-l}y - j) \tag{3.5}$$

and the type M can take three values (01, 10 or 11) corresponding to one of the three mother wavelets for horizontal, vertical and diagonal differencing.

Each t-uplet (s,i,j) defines a *wavelet square* at scale s and an offset (i,j). Thanks to the Haar model property, the projection of the occupancy function over a basis vector function Φ^{sij} is given by the following scalar product of Φ^{sij} with the occupancy function:

$$< p(x,y) \mid \Phi^{sij} >= \int_{x,y \in \mathbb{R}^2} p(x,y)\Phi^{sij}(x,y)d_{x,y} \tag{3.6}$$

Logarithmic form for OG updates

In the sequel, we will omit the index (x,y) associated to the OG cells (for simplifying the notations). A standard update in OG is defined by the following equation:

$$p([O^t = occ]) = \frac{p([O^{t-1} = occ])p(z^t|[O^{t-1} = occ])}{\sum_{o^t} p(o^t)p(z^t|o^t)} \tag{3.7}$$

Let $p^t(occ) = p([O^t = occ])$ and $p^t(emp) = p([O^t = emp])$; then we can write $p^t(occ) = 1 - p^t(emp)$, and we can summarize this property using a single coefficient q^t:

$$q^t = p^t(occ)/p^t(emp) \tag{3.8}$$

Substituting eq. 3.7 into eq. 3.8 leads to the elimination of the marginalisation term; then, we can write the following recursive formulation:

$$q^t = \frac{p(z^{t-1}|occ)}{p(z^{t-1}|emp)}q^{t-1} = q^0 \prod_{i=0}^{t-1} \frac{p(z^i|occ)}{p(z^i|emp)} \tag{3.9}$$

where q_{t-1} is recursively evaluated.

Such updating operations are clearly non-linear. In order to be able to perform the update operations using only sums, we have decided to make use of a logarithmic form of the model:

$$odds^t = odds^0 + \sum_{i=0}^{t-1} \log(\frac{p(z^i|occ)}{p(z^i|emp)}) \tag{3.10}$$

where $odds^t = log(q^t)$.

Let $Obs(z^{t-1})$ be the term $\log(\frac{p(z^{t-1}|occ)}{p(z^{t-1}|emp)})$ in the eq. (3.10). $Obs(z^{t-1})$ represents the "observation" term of the update, and $odds^(t-1)$ is the a priori

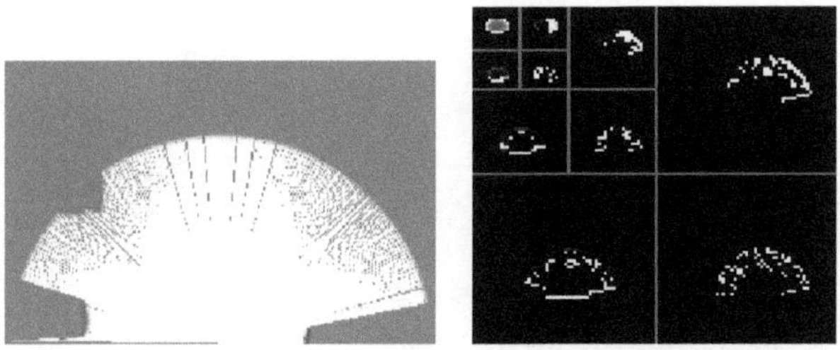

(a) (b)

Fig. 3.6. Mapping obtained by a laser range-finder : (a) the obtained OG model; (b) the related three first "detail spaces" of the WOG model and the complementary scaled space. The density of non-zero coefficients decreases drastically in wavelet space.

term. Then, updating a WOG can be done by adding the wavelet transform of observation terms to the wavelet representation of the map.

Multi-scale semantic of logarithmic OG

In the case of the Haar wavelet basis (eq. 3.5), $\Phi_s^{(i,j)}$ has a constant value (k) over the domain[3] (s, i, j), and is equal to zero elsewhere. Then the integral of the scalar product is a sum over the cells enclosed in the domain (s, i, j); this sum can be re-written as follows [16]:

$$< \log(\frac{p^t(x,y)}{1-p^t(x,y)}) \mid \Phi^{sij} >$$

$$= \int_{x,y \in \mathbb{R}^2} \log(\frac{p^t(x,y)}{1-p^t(x,y)}) \Phi^{sij}(x,y) d_{x,y} \tag{3.11}$$

$$= k \log(\frac{\Pi_{c \in (s,i,j)} p([O_c = occ])}{\Pi_{c \in (s,i,j)} p([O_c = emp])}) \tag{3.12}$$

$$= k \log(q_s) \tag{3.13}$$

where c is the index of a cell in the square domain. $k \log(q_s)$ is the log of the ratio of two geometric means (cells are "all occupied" and cells are "all empty") which leads us to a continuous semantics. Let define *full* as the event "every subcell is occupied" then:

$$\Pi_{c \in (s,i,j)} p([O_c = occ]) = p(\bigwedge_{c \in (s,i,j)} [O_c = occ]) = p(full)$$

Let define *open* as the event "every subcell is empty".

q_s leads immediately to the conditional probabilities:

$$P(open | open \vee full) = q_s/(1+q_s) \tag{3.14}$$

$$P(full | open \vee full) = 1/(1+q_s) \tag{3.15}$$

which express a continuous ratio between the two events *full* and *open* (Fig. 3.7).

[3] With value: $k = 2^{-s}$.

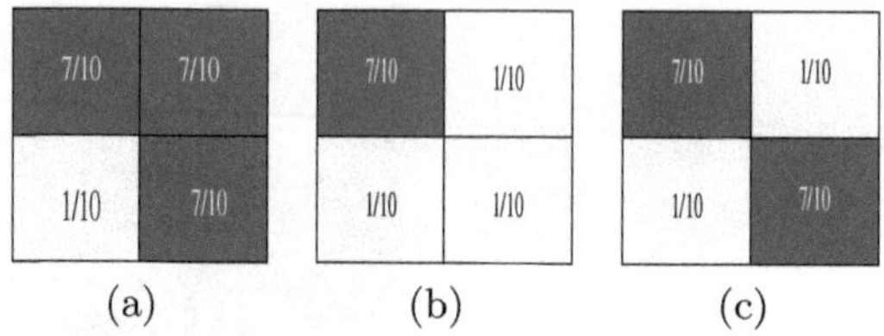

Fig. 3.7. Three configurations of a subsquare domain at scale 1/2: the probability that the square domain 7(a) is fully "occupied" is 0.0343; its probability of being fully "empty" is 0.0243. The occupancy ratio related to the "occupied" and "empty" properties is $P(open|open \vee full) = 0.58$; this means that the related square domain is considered as "occupied". The occupancy ratio in 7(b) is 0.003; it is of 0.06 in 7(c).

The multi-scale information which can be computed by this approach, is directly derived from these two basic cases. Consequently, only 2 relevant events can be considered for a square domain (s, i, j) containing n subcells (i.e. where 2^n possibilities can be defined by the binary vector representing the possible occupancy of these n subcells), Fig. 8(c) illustrates. Fig. 3.7 shows the information which can be derived from the previous multi-scale occupancy model. The next step will be to exploit this concise information for elaborating multi-scale algorithms.

(iii) WOG implementation and experimental results

This approach has been implemented and tested on the Cycab equipped with a laser sensor an with our SLAM method [17]. In the experiments, the size of the finer scale is equal to $6.25cm$, and we have used 5 different scales (where the size is multiplied by 2 for each new scale); thus, the size of the coarsest cells is equal to 1 m. The size of the square window which is used for constructing the WOG model, is chosen in order to contain the whole sensor field of view. The content of this window is updated at each laser scan (see Fig. 3.6; then, the Haar wavelet transform is applied to this sensory data, and incrementally added to the current WOG model. The chosen compression threshold is a compromise between data fitting and sparsity (wavelet terms which are lower than the threshold are removed).

Fig. 3.8 shows the results obtained on the car park of INRIA. These experimental results clearly shows that we have obtained a significant reduction of the size of the model (about 80% relatively to the OG model), and that the interesting details are still represented (such as the beacons represented by dark dots in Fig. 8(b)). It should be noticed that the coarser model give a quite good representation of the empty space (see Fig. 8(c)); this model could be used for path planning, and refined when necessary using the wavelet model. In the previous experiments, the map building has been done in real-time.

One of the major issue of our approach is now, to combine bayesian occupancy filter and object tracking such as the informations about the object motions

(a) (b) (c)

Fig. 3.8. Maps of the car park of INRIA. (a) The *OG* model contains 393, 126 cells, while the *WOG* model contains 78, 742 cells. (b) The *OG* model reconstructed from the previous *WOG* model; it can be seen that significant details such as beacons have been captured by the *WOG* representation (the shapes of maps (a) and (b) are very close). (c) The empty space is pretty well approximated at the coarser scale.

could be transmitted to the high level object motion prediction module. One of the possibilities, we explore now is to extract coherent zones of the BOF in term of occupancy and velocity and that we call them objects. Our hope is that these zones would appear hierarchically through the scales in the wavelet representation, and so woubld be easy to find.

3.3 Medium-Term Motion Prediction

Motion planning for dynamic environments is a very active research domain. Because the problem is NP-Hard [18], most of the research effort has been directed towards finding algorithms that are able to cope with this complexity. There is, however, another aspect of motion planning that is often overlooked despite its importance: motion planning algorithms need to know in advance the motion of the objects which populate the environment.

The problem is that, in most real applications, the future motion of the moving objects is a priori unknown, making it necessary to predict it on the basis of observations of the objects' past and present states. These observations are gathered using various sensors (*eg* radars, vision systems, etc.) which have limited precision and accuracy.

Until recently, most motion prediction techniques have been based on kinematic or dynamic models that describe how the state (*eg* position and velocity) of an object evolves over time when it is subject to a given control (*eg* acceleration) (cf. [19]). These approaches proceed by estimating the state, using techniques such as the Kalman Filter [20], and then applying the estimate to its motion equations in order to get state predictions.

Although these techniques are able to produce very good short-term predictions, their performance degrades quickly as they try to see further away in the future. This is especially true for humans, vehicles, robots, animals and the like, which are able to modify their trajectory according to factors (*eg* perception, internal state, intentions, etc.) which are not described by their kinematic or dynamic properties.

To address this issue, a different family of approaches has emerged recently. It is based on the idea that, for a given area, moving objects tend to follow typical motion patterns that depend on the objects' nature and the structure of the environment. Such approaches operate in two stages:

1. *Learning stage:* observe the moving objects in the workspace in order to determine the typical motion patterns.
2. *Prediction stage:* use the learnt typical motion patterns to predict the future motion of a given object.

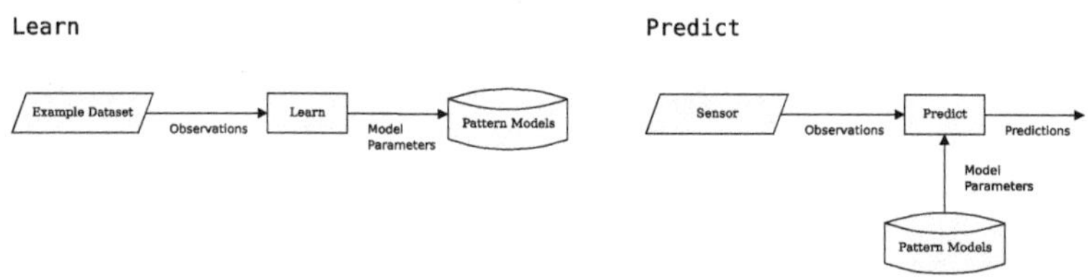

Fig. 3.9. Learn then predict

Thus, learning consists in observing a given environment in order to construct a representation of every possible motion pattern. But, how long should we observe the environment in order to construct such a "pattern library"? Given the enormous number of possible patterns for all but the simplest environments, there is not a simple answer. This raises an important problem of existing learning techniques [21, 22, 23]: they use a "learn then predict" approach, meaning that the system goes through a learning stage where it is presented with a set of observations (an example dataset) from which it builds its pattern models. Then, the models are "frozen" and the system goes into the prediction stage.

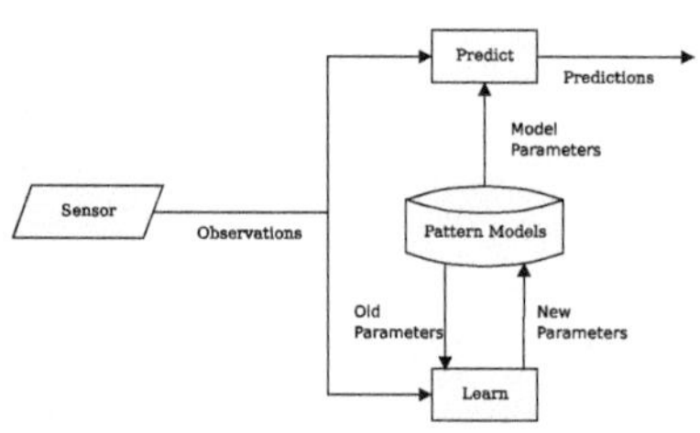

Fig. 3.10. Learn and Predict

The problem with this approach is that it makes the implicit assumption that all possible motion patterns are included in the example dataset, which, as we have shown, is a difficult condition to meet. In this paper we present an approach which, in contrast to the approaches mentioned above , works in a "learn and predict" fashion (fig. 3.10). That is, learning is an incremental process, which continuously refines knowledge on the basis of new observations which are also used for prediction. To the extent of our knowledge, this is the first learning based motion prediction technique in the literature to have this property.

Our work is based on Hidden Markov Models [24], a probabilistic framework used to describe dynamic systems which has been successfully applied to "learn then predict" approaches [22, 23]. A Hidden Markov Model (HMM) may be viewed as a graph, where nodes represent states (*eg* places in the environment) and edges represent the transition probability of moving from one node to another in a single time step, the number of nodes and the existence of nodes determines the model's structure. The model also assumes that states are not directly observable but produce observations with a given probability.

Our approach is founded on the hypothesis that objects move in order to reach specific places of the environment called goals. Hence, motion patterns are represented using a different Hidden Markov Model (HMM) for every goal. Each such HMM describes how objects move to reach the corresponding goal. It is this set of HMMs that are used for prediction purposes.

Our main contribution is a novel approximate HMM learning technique based on the Growing Neural Gas algorithm [25]. It is able to process observations incrementally (*ie* one by one), in order to find the HMM's structure as well as to estimate the model's transition probabilities. Moreover, the algorithm is adaptive, meaning that it is able to insert or delete states or transitions in order to respond to changes in the underlying phenomenon. The same incrementallity and adaptivity properties apply to goal identification, hence, it is able to create models for motion patterns that have just been discovered or to delete old ones when the corresponding patterns are no longer observed. The cost of learning for each iteration is linear with respect to the number of states, as is the inference (*ie* prediction). This enables real-time processing, which is indispensable for a "learn and predict" approach.

3.3.1 Theoretical Framework

This section provides a gentle introduction to the two main tools which are used by our approach. Readers which are already familiar with the Hidden Markov Models or the Growin Neural Gas algorithm may safely skip the corresponding sections and proceed directly to §3.3.2.

Hidden Markov Models

This is a concise discussion on Hidden Markov Models (HMM), the interested reader is referred to [24] for an excellent tutorial on the subject.

An HMM may be viewed as a stochastic automaton which describes the temporal evolution of a process through a finite number of discrete states. The process progresses in discrete time steps, going from one state into another according to a given *transition probability*. It is assumed that the current state of the system is not directly observable, instead, it produces an output (i.e. observation) with a certain probability known as the *observation probability*.

Representation

An HMM describes the system using three stochastic variables: a) the present state s_t, b) the previous state s_{t-1}, and the current observation o_t, the rest of the model is defined by the following elements:

- The number of discrete states in the model N. A discrete state is denoted by its number $i : 1 \leq i \leq N$.
- The transition probability function, expressed by $P(s_t \mid s_{t-1})$. This probability is represented with a $N \times N$ *transition matrix* A where each element $a_{i,j}$ represents the probability of reaching the state j in the next time step given that the system was in state i.

$$a_{i,j} = P([s_t = j] \mid [s_{t-1} = i]) \tag{3.16}$$

- The observation probability function, expressed by $P(o_t \mid s_t)$. In general, for each state, the observation probability is represented by a gaussian distribution[4].

$$P(o_t \mid [s_t = i]) = \mathbb{G}(o_t; \mu_i, \sigma_i) \tag{3.17}$$

  The set of all the gaussians' parameters is denoted by $B = \{(\mu_1, \sigma_1), \cdots, (\mu_N, \sigma_N)\}$.
- The initial state distribution, expressed by $P(s_0)$. This represents our knowledge about the state of the system before receiving any observation and is denoted by Π. It is often represented by a uniform distribution $P([s_0 = i]) = \frac{1}{N}$ or by a $N \times 1$ matrix where $P([s_0 = i]) = \Pi_i$.

The three probability functions defined above form a Joint Probability Function (JPD) which encodes two conditional independence assumptions: a) knowing the state, observations do not depend on each other, and b) knowing the present state, the past and future states are mutually independent (eq. 3.18).

$$P(s_{t-1}, s_t, o_t) = P(s_{t-1})P(s_t \mid s_{t-1})P(o_t \mid s_t) \tag{3.18}$$

The parameters of these three probabilities are often denoted using the compact notation $\lambda = \{A, B, \Pi\}$, they are known together as the *model's parameters*, as opposed to the *model's structure*, which we will now discuss.

The structure of a model, Ξ, is the specification of its number of states N and the valid transitions between states. It may be visualized using the model's connectivity graph, where each vertex represents one state, vertices are joined by directed edges such that every strictly positive element of A ($a_{i,j} > 0$) will be represented by a directed edge from vertex i to vertex j. Hence, a matrix having only strictly positive elements corresponds to a fully connected, or *ergodic*, structure graph, while a sparse matrix corresponds to a graph having fewer edges. This is illustrated in fig. 3.11.

[4] It is also common to represent it by a mixture of gaussians, or, for discrete observations, by a table.

Structure is important because it affects the complexity of inference in the model, for example, filtering (state estimation) has complexity $O(N^2)$ for a fully connected structure graph, however, if the graph is sparse and each state has at most P predecessors, the complexity is $O(NP)$ [26]. Moreover, structure also influences the quality of inference [27, 28].

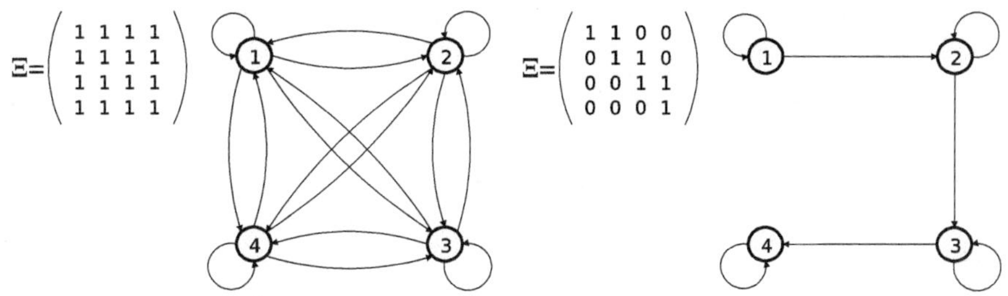

(a) $O(N^2)$ edges (b) $O(N)$ edges

Fig. 3.11. Two examples of an order 4 structure graph

Inference

HMM's are used to perform bayesian inference: compute the probability distributions of unknown variables given the known ones. Like in other Dynamic Bayesian Networks, inference is often used in HMM's to perform filtering $P(s_t \mid o_{1:t})$, smoothing $P(s_t \mid o_{1:T})$ and prediction $P(s_{t+K} \mid o_{1:t})$. Other two probabilistic questions which are more specific to HMM's are: a) evaluating the probability of a sequence of observations, and b) finding the sequence of states that is more likely to correspond to a sequence of observations. This last question is answered using a dynamic programming technique known as the Viterbi Algorithm [29].

Learning

In order to perform inference with an HMM, it is necessary to define its structure Ξ as well as its parameters λ. But, how to choose these values for a particular application? The solution is to estimate (*ie* learn) these values from data.

Most approaches in the literature assume that the structure is known and only try to learn the model's parameters λ. The most widely used of these *parameter learning* approaches is the Baum-Welch algorithm which is an application of Expectation-Maximization [30].

Despite being an active research subject, *structure learning*, no structure learning algorithm has become standard, probably the most popular are [31] and [32].

Growing Neural Gas

We will briefly introduce the Growing Neural Gas (GNG) algorithm, which is explained in detail in [25]. The Growing Neural Gas is an unsupervised competitive learning algorithm which may be applied to a variety of problems (e.g. vector quantization, topology learning, pattern classification and clustering). It processes input vectors and constructs a network of N elements called *units*, each of these units has an associated D-dimensional vector called its *weight* (w_i) and is linked to other units called its *neighbors* $\mathfrak{N}_i$. The algorithm starts with two units linked together, then, as new input vectors are processed, units and links may be added or deleted to the network.

The algorithm implicitly partitions the whole space into N *Voronoi regions* V_i which are defined by:

$$V_i = \{x \in \mathbb{R}^D : \|x - u_i\| \leq \|x - u_j\|, \forall j \neq i\} \quad (3.19)$$

We illustrate this in 2-dimensional space (fig. 3.12). Each unit occupies its own Voronoi region, given an input vector, the winning unit is the one having its weight vector in the same Voronoi region.

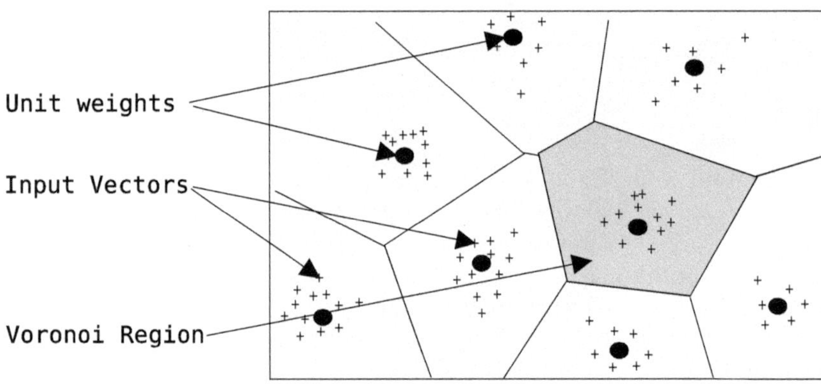

Fig. 3.12. Implicit partition: there are some 2-dimensional input vectors (crosses). The units' weights are represented by points and Voronoi regions are separated by boundary lines.

During learning, the units' weights are modified in order to minimize the *distortion* which is the mean distance between the winners and their corresponding input vectors. The algorithms also builds incrementally a topology of links which is a subset of the Delaunay triangulation of the units' weight vectors. This means that, in order to be linked together, two units must have a common border in the Voronoi region (see fig. 3.13).

Recapitulating, the GNG algorithm has the following properties:

1. No previous knowledge about the number of units N is required.
2. It is incremental: the model is updated by processing input in a one by one basis.
3. It is adaptive: units and / or links may be added or deleted to reflect changes in the underlying phenomenon.

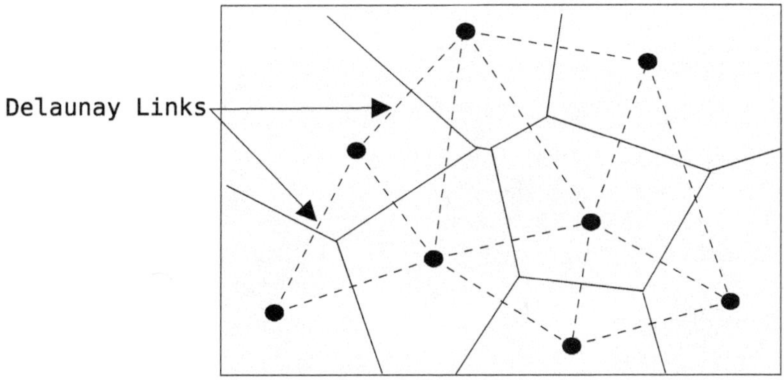

Fig. 3.13. The Delaunay triangulation for the previous example. Delaunay links are represented by dashed lines. Notice how the number of links corresponds to the number of borders for a given area.

4. It minimizes the distorion or quantization error: units' weights are placed in order to minimize their distance to input vectors.
5. Links are a subset of the Delaunay triangulation: they connect neighbor Voronoi cells.

3.3.2 Proposed Approach

The approach we propose consists of an algorithm which is able to continuously predict future motion based on a model which is constantly improved on the basis of observed motion. This *learn and predict* capability constitutes an advantage over existing techniques, making it unnecessary to have a learning dataset containing at least one example of every observable motion pattern.

The input of our algorithm consists of position observations $o_t = (x_t, y_t)$ and a trajectory termination flag η_t which is set to one when the observation corresponds to the end of a trajectory (*ie* when the object has stopped moving for a long time or exited the environment) and is set to zero otherwise. At every time step, this input is used to compute a probabilistic estimate of the state with a lookahead of K timesteps ($P(s_{t+k} \mid o_t)$) as well as to update the model (fig. 3.10).

Representation

Our approach is based on the hypothesis that objects move in order to reach specific places in the environment (*ie* goals). The idea is to identify all the possible goals. Then, for each of these goals, we construct a Hidden Markov Model (*ie* motion model) representing how an object should move in order to reach it.

It is assumed that transition probabilities depend on the goal to be reached (denoted by γ hereafter) and that structure and observation probabilities are independent of the goals, hence, they are the same for all motion models.

These assumptions lead to the following JPD:

$$P(s_{t-1}, s_t, o_t, \gamma) = P(s_{t-1})P(\gamma)P(s_t \mid s_{t-1}, \gamma)P(o_t \mid s_t) \tag{3.20}$$

So our model is defined by the following:

- The number of goals in the model G.
- The number of states in the model N.
- The transition probability function $P(s_t \mid \gamma, s_{t-1})$. This probability is represented with an unnormalized $G \times N \times N$ transition matrix[5] A where every element $a_{g,i,j}$ represents the number of times that a transition from state i to state j has been observed, given that the object is going to goal g.

$$a_{g,i,j} = P([s_t = j] \mid [\gamma = g][s_{t-1} = i]) \tag{3.21}$$

- The observation probability function $P(o_t \mid s_t)$. Is represented by gaussian distributions. The set of all the gaussians' parameters is denoted by $B = \{(\mu_1, \sigma_1), \cdots, (\mu_N, \sigma_N)\}$.
- The initial goal distribution $P(\gamma_0)$. Is considered to be a uniform distribution $P([s_0 = i]) = \frac{1}{G}, \forall i \in [1, N]$.
- The initial state distribution $P(s_0)$. Is considered to be a uniform distribution $P([s_0 = i]) = \frac{1}{N}, \forall i \in [1, N]$.

The set of *pattern models* shown in fig. 3.10 consists of structure and parameters. The structure is defined by the number of goals G, the number of states N, and the structure graph Ξ which is the same for all the A_g matrices.

The parameters consists of the transition table A, and the observation parameters B. Given that both $P(s_0)$ and $P(\gamma)$ are considered to be uniform, they do not require any extra parameter.

Learning

Learning is composed of several subprocesses, an overview is presented in fig. 3.14.

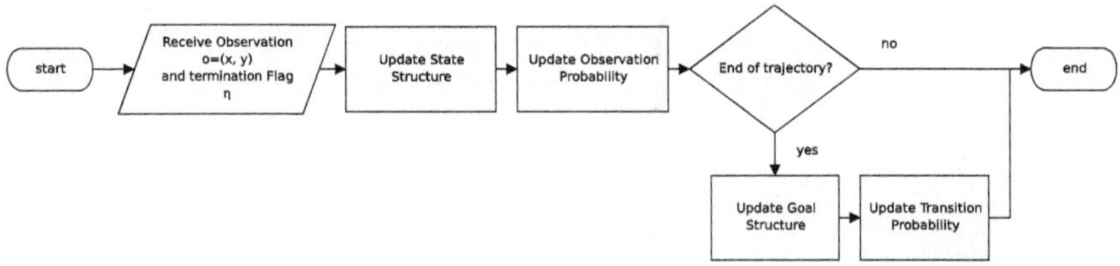

Fig. 3.14. Learning Overview

Updating State Structure

The input observation o_t is passed as an input vector to a GNG network $\mathfrak{G}_{state}$. Nodes of this network represent states, and links represent allowed transitions

[5] This is considered equivalent to having G $N \times N$ matrices $A_g : 1 \leq g \leq G$.

(*ie* each link represents two transitions, corresponding to the two possible directions). This network is used to update A as follows:

1. If a new unit has been inserted or deleted in $\mathfrak{G}_{state}$, a corresponding row and column are inserted to or deleted from all A_g matrices.
2. If a new link has been added or deleted from $\mathfrak{G}_{state}$, two corresponding transitions are added to or deleted from all A_g.

Updating Observation Probabilities

The observation probabilities are directly computed from $\mathfrak{G}_{state}$. Unit weights are used as mean values for the gaussians' centers:

$$\mu_i = w_i, \forall w_i \in \mathfrak{G}_{state} \tag{3.22}$$

And σ_i is estimated from the average distance to unit's neighbors:

$$\sigma_i = \frac{C}{|\mathfrak{N}_i|} \sum_{j \in \mathfrak{N}_i} \|w_i - w_j\| \tag{3.23}$$

Where C is a weighting constant, in our experiments, we have set it to 0.5.

Updating Goal Structure

Goals are discovered by clustering together observations that correspond to trajectories endpoints, which is indicated by $\eta = 1$. Given that the number of goals is ignored, we will use another GNG structure $\mathfrak{G}_{goal}$ to perform the clustering. The number of clusters G corresponds to the number of units in $\mathfrak{G}_{goal}$. This means that, when a new unit is inserted or deleted in $\mathfrak{G}_{goal}$ a corresponding slice (*ie* matrix) is inserted to or deleted from A.

Updating Transition Probabilities

In this step, the entire observation sequence $\{o_1, \cdots, o_T\}$ for a trajectory is used, this means that it is necessary to store all past observations in an internal data structure.

In order to choose the transition probability table to update, the attained goal γ should be identified, this is done by choosing the goal which is closest to the last observation.

$$\gamma = \arg\min_i \|o_T - w_i\|, \forall w_i \in \mathfrak{G}_{goal} \tag{3.24}$$

Transition probabilities are then updated by applying maximum likelihood: we use the Viterbi algorithm to find the most likely sequence of states $\{s_1, \cdots, s_T\}$ and then, we increment the corresponding counters in A[6].

$$a_{\gamma,s_{t-1},s_t} = a_{\gamma,s_{t-1},s_t} + 1, \forall t : 1 \leq t \leq T-1 \tag{3.25}$$

[6] This contrasts from Baum-Welch which performs weighted counting according to the likelihood of every *possible* state sequence.

A problem with this approach is that the transition counters of new links in the topology will have a much smaller value than those that correspond to older links, hence, old links will dominate. This is solved by multiplying transition weights by a *fading factor* f. In general this is similar to having a bounded counter with a maximum value of $\frac{1}{1-f}$.

$$a_{\gamma,s_t,i} = a_{\gamma,s_t,i} \times f, \forall t, i : 1 \leq t \leq T-1, i \in \mathfrak{N}_{s_t} \tag{3.26}$$

Instead of normalizing A, normalization is performed when computing the transition probability.

$$P([s_t = j] \mid [\gamma = g][s_{t-1} = i]) = \frac{a_{g,i,j}}{\sum_{k \in \mathfrak{N}_i} a_{g,i,k}} \tag{3.27}$$

Finally, the data structure that was used to store the observation sequence may be cleared.

Prediction

Prediction is performed in two steps. First, the belief state (*ie* the joint probability of the present state and goal) is updated using the input observation o_t:

$$P(\gamma, s_t \mid o_t) = \frac{1}{Z} P(o_t \mid s_t) \sum_{s_{t-1}} P(s_t \mid \gamma, s_{t-1}) P(\gamma, s_{t-1} \mid o_{t-1}) \tag{3.28}$$

Where $P(\gamma, s_{t-1} \mid o_{t-1})$ is the belief state calculated in the previous time step.

Then, motion state is predicted with a look-ahead of K time steps using:

$$P(\gamma, s_{t+K} \mid o_t) = \sum_{s_{t+K-1}} P(s_{t+K} \mid \gamma, s_{t+K-1}) P(\gamma, s_{t+K-1} \mid o_t) \tag{3.29}$$

Finally, the state is obtained by marginalizing over the goal:

$$P(s_{t+K} \mid o_t) = \sum_{\gamma} P(\gamma, s_{t+K} \mid o_t) \tag{3.30}$$

An alternative to state prediction is to predict the goal by marginalizing the state from the belief state:

$$P(\gamma \mid o_t) = \sum_{s_t} P(\gamma, s_t \mid o_t). \tag{3.31}$$

3.3.3 Experimental Results

Test Environment

The environment we have chosen to validate our approach is the main lobby of our research institute. It features the main entrance to the building, a self-information directory post, the front-desk, a cafeteria area and a number of doors

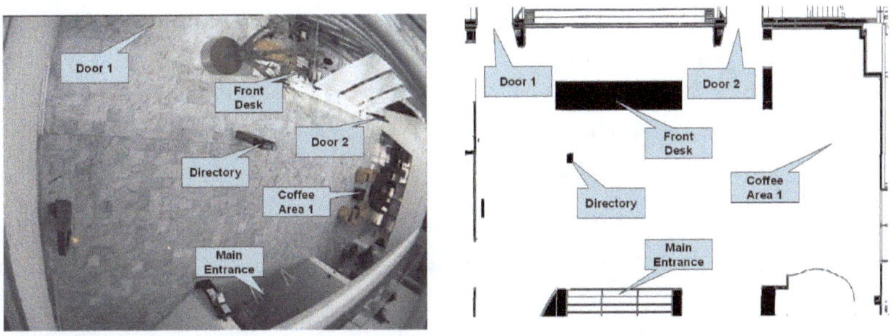

Fig. 3.15. Inria's main lobby: video view (left) and 2D map (right)

leading to various halls, rooms and auditoriums (Fig. 3.15). This environment is the heart of the institute, all the personnel passes through it at some point during the day for a reason or another (going in or out, coffee break, attending a lecture, etc). This environment is interesting because the motion patterns of the people is rich and the flow of people sufficient to ensure that it will be possible to gather a significant number of observations.

The testing of our approach has been done in two stages. First, we have used observation data coming from a software simulating the trajectories of people in the main lobby. Then we have used live observations coming from a visual tracking system.

The interest of using simulated data is that it allows to evaluate our approach in controlled conditions for which it is possible for instance to predefine the number of motion patterns.

In both cases, we have gathered a significant number of observations. We have presented 1000 trajectories to the learning algorithms in order to build an initial model, before starting to measure the results. Then, prediction performed has been measured using another 300 trajectories. The test results obtained with simulated data (resp. real data) are presented in section 3.3.3 (resp. 3.3.3).

Simulated Observations

Getting the Observations

The simulating system we have developed relies upon a number of control points representing "'places of interest" of the environment such as the doors, the front-desk, etc. Based upon this set of control points, a set of 32 typical motions patterns has been defined. Each motion pattern consists in a sequence of control points to be traversed. An observed trajectory is computed in the following way: first, a motion pattern is randomly chosen. This motion pattern provides a set of control points. Then, *goal points* corresponding to each of these control points are randomly generated using two-dimensional Gaussian distributions whose mean value are the control points. Finally, the motion between two goal points in the sequence is simulated using discrete, even-size steps in a direction drawn from a Gaussian distribution whose mean value is the direction of the next goal point.

Switching from one goal point to the next is done when the distance to the current goal point is below a predefined threshold.

We have generated the 1300 trajectories that were required for our experiments. an image of trajectories in the data set are presented in fig. 3.16.

Learning Results

We have run our algorithm against the simulated data set 16(a). The algorithm took about 2 minutes to process the 1000 initial trajectories, which contained a total of 58,262 observations meaning an average processing frame rate of about 480 observations per second. As a result of the learning process, a structure having 196 states has been found (figs. 16(b) and 16(c)). Also a total of eight goals were identified, as it may be seen in fig. 16(d) the detected goals seem to correspond to many of the interesting points shown in fig. 3.15.

Prediction Results

Figure 3.17 illustrates the prediction process. The figures show the real future trajectory of an object, which is unknown for the system. The figures show the probability of reaching a particular goal (indicated by the size of the cubes) as well as the state probability in $t + 3$ seconds. For visual purposes this has been illustrated with particles, where a higher concentration of particles indicates a higher state probability.

In order to test prediction performance, we measure the distance between the predicted goal and the real final destination of the object (fig. 3.18). For each of the 300 trajectories we measure this distance when 10% of the total trajectory has been seen, then, we do the same for 20% of the total trajectory and so on until 90%.

We think that the obtained results are quite good. Even when only 30% of the total trajectory is used to predict, the mean error is of about 3 meters which

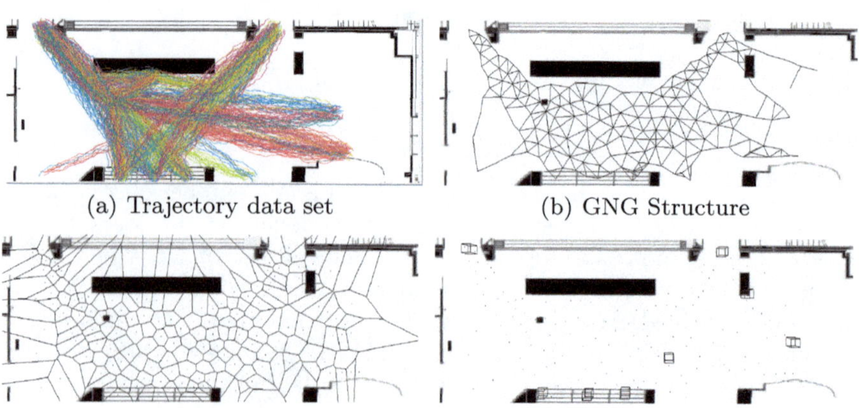

(a) Trajectory data set
(b) GNG Structure
(c) Voronoi Diagram
(d) Learned Goals

Fig. 3.16. Overview of the learning process (simulated data)

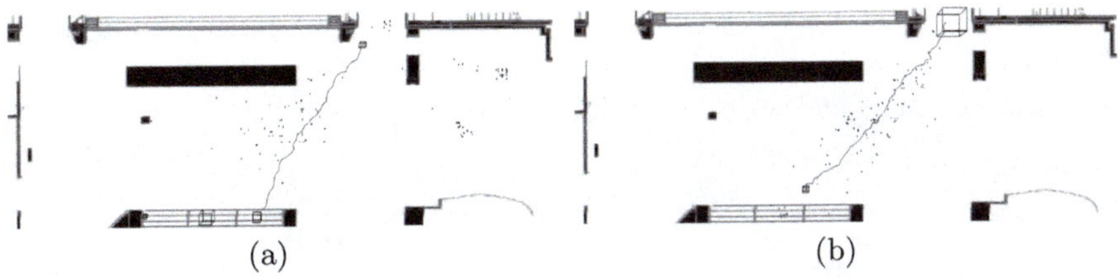

Fig. 3.17. Prediction Examples (simulated data)

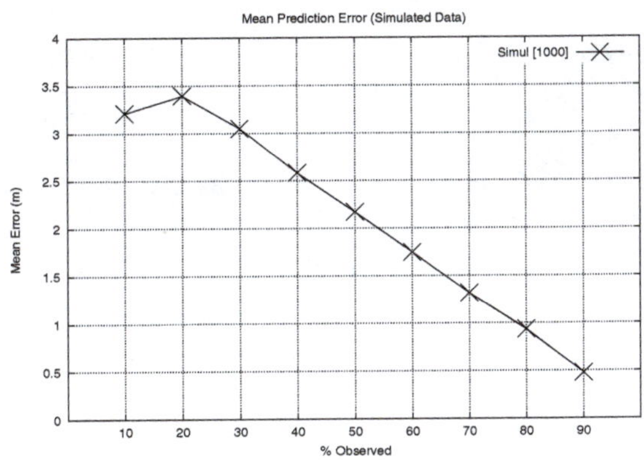

Fig. 3.18. Prediction Error (simulated data)

is relatively low with respect to the size of the environment and the distance between goals.

Real Observations

Getting the observations

To gather observations about the motions performed in the test environment, we use the visual tracking system proposed in [33]. This system detects and tracks objects moving in the images of a video stream. The information collected (position and size of the moving object, etc.) is then projected into the 2D map

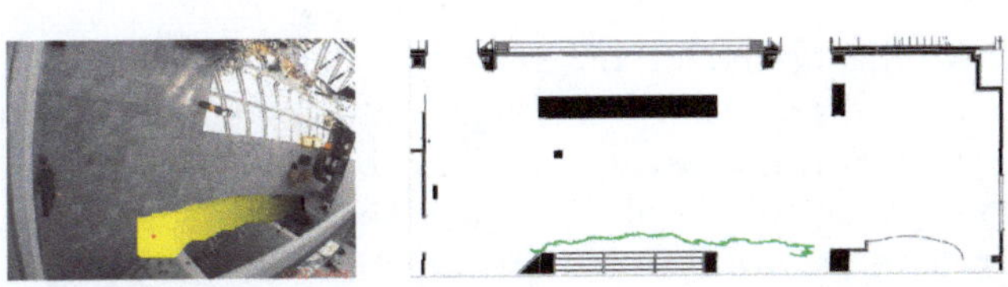

Fig. 3.19. Architecture of the visual tracking system (top). Example of a motion observed in the image (bottom-left), and its projection in the 2D map of the environment (bottom-right).

of the environment. The data flow of the overall tracking system features a detector-tracker, a module to correct the distortion of the video camera, and a final module to project the information in the 2D map of the floor.

Learning Results

Figure 20(a) shows the real data set that we have obtained. It is important to notice that trajectories in the data set do not cover the entire environment. This happens because of clipping due to the distortion correction algorithm we have used.

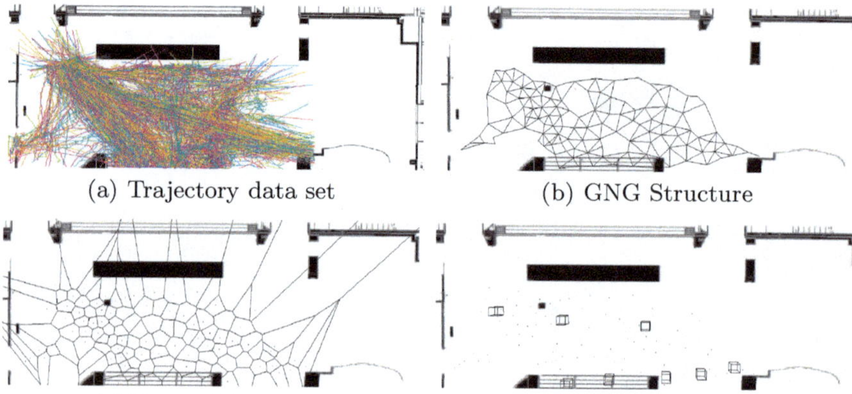

(a) Trajectory data set (b) GNG Structure

(c) Voronoi Diagram (d) Learned Goals

Fig. 3.20. Overview of the learning process (real data)

The algorithm took about 70 seconds to process the 36,444 observations of the first 1000 trajectories in the data set for an average of about 520 observations per second. The found structure (figs. 20(b) and 20(c)) consisted of 123 states. The algorithm has detected eight goals, shown in fig. 20(d). In this case, some of the goals do not correspond to interesting points of the environments. This happens because of two factors: a) clipping reduced the environment an creates some fake entry and exit points, and b) the tracking system sometimes losses the identifier of an object resulting in single trajectories being broken down in several smaller trajectories.

Prediction Results

Figure 3.21 illustrates prediction based on real data. The example in the right is especially interesting because, although state prediction does not seem to correspond to the real trajectory, the goal has been correctly predicted.

In fig. 3.22 we present the results of our prediction performance measure (see §3.3.3). Here, the results are even better than for simulated data, which may be surprising at first. The reason is again clipping, which reduces the effective size of the environment, thus bringing goals closer together. In the other hand, the noisy nature of real data and is reflected by the slower convergence of real data when compared with simulated one.

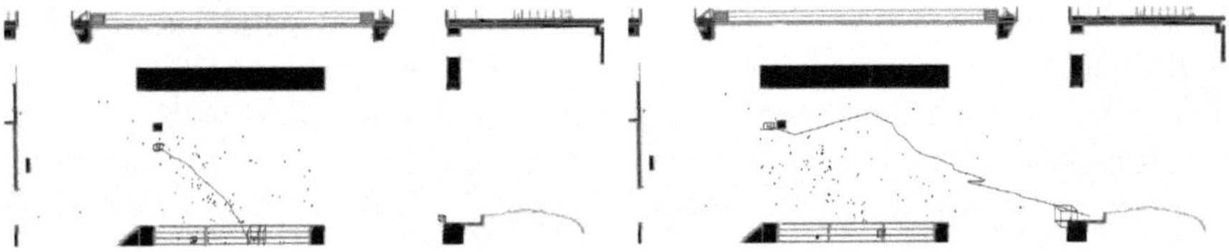

(a) (b)

Fig. 3.21. Prediction Examples (real data)

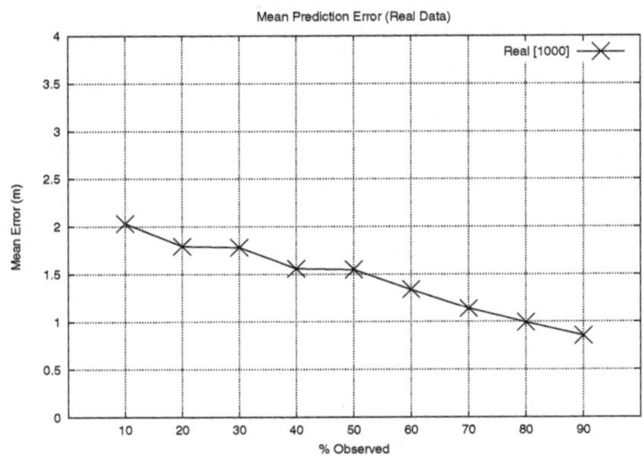

Fig. 3.22. Prediction Error (real data)

3.4 Safe Navigation in Dynamic Environments

When placed in a dynamic environment, an autonomous system must consider the real time constraint that such an environment imposes. Indeed, the system has a limited time only to make a decision about its future course of action otherwise it might be in danger by the sole fact of being passive. The time available depends upon a factor that we will call the *dynamicity* of the environment and which is a function of the system and the moving objects'dynamics.

In this context, basing the decision making process on a motion planning technique[7] leaves little hope to fulfil this real-time constraint given the intrinsic time complexity of the motion planning problem [34] (even if using randomised approaches). This certainly explain why so many reactive methods[8] have been developed in the past (*cf* [35], [36], [37], [38], [39], [40], [41], [42], [43], [44] or [45]. However, reactive approaches are confronted with two key issues: the *convergence* and the *safety* issues. As for convergence, their lack of lookahead may prevent the system to ever reach its goal. As for safety, what guarantee is there that the system will never find itself in a dangerous situation eventually yielding to a collision?

[7] Wherein a complete motion to the goal is computed a priori.

[8] Wherein only the next action is determined at each time step.

Partial Motion Planning (PMP) is the answer we propose to the problem of navigation in dynamic environments. It is especially designed in order to take into account the real-time constraint mentioned above. PMP is a motion planning scheme with an anytime flavor: when the time available is over, PMP returns the best partial motion to the goal computed so far. Like reactive scheme, PMP is also confronted to the convergence and safety issues. At this point, we have decided to focus on the safety issue and to propose a solution relying upon the the concept of *Inevitable Collision States* (ICS) originally introduced in [46]. An ICS is a state such that no matter what the future motion of the system is, it eventually collides with an obstacle. ICS takes into account the dynamics of both the system and the moving obstacles. By computing ICS-free partial motion, the system safety can be guaranteed.

PMP is detailed in section 3.4.1 while section 3.4.2 presents the ICS concept. Finally, an application of PMP to the case of a car-like system in a dynamic environment is presented in section 3.4.3.

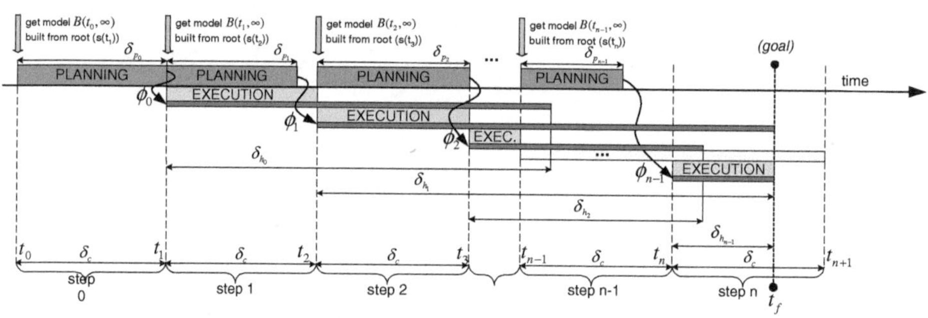

Fig. 3.23. Partial Motion Planning iterative cycle

3.4.1 Partial Motion Planning

As mentioned earlier, a robotic system cannot in general safely stand still in a dynamic environment (it might be collided by a moving obstacle). It has to plan a motion within a bounded time and then execute it in order to remain safe. The time δ_c available to calculate a new motion is function of the nature and dynamicity of the environment. To take into account the real-time constraint that stems from a dynamic environment, we propose a scheme that calculates partial motions only according to the the following cycle (also depicted in Fig. 3.23):

PMP Algorithm	
Step1:	*Get model of the future*
Step2:	*Build tree of partial motions towards the goal*
Step3:	*When δ_c is over, return best partial motion*
Step4:	*Repeat until goal is reached*

Like motion planning, partial motion planning requires a model of the environment, the first step is aimed at getting this model. The required model is provided by the environment modelling and motion prediction functions presented earlier. The periodic iterative PMP scheme proposed in this paper accounts for both the planning time constraints and the validity duration of the predictions made.

3.4.2 Inevitable Collision States

Like every method that computes partial motion only, PMP has to face a safety issue: since PMP has no control over the duration of the partial trajectory that is computed, what guarantee do we have that the robot $\mathcal{A}$ will never end up in a critical situation yielding an inevitable collision? As per [46], an Inevitable Collision State (ICS) is defined as a state s for which no matter the control applied to the system is, there is no trajectory for which the system can avoid a collision in the future. The answer we propose to the safety problem lies then in the very fact that the partial trajectory that is computed is ICS-free. Meaning that, even in the worst case scenario where the duration δ_{h_i} of the partial trajectory is shorter than the cycle time δ_c, $\mathcal{A}$ can always execute one of the existing safe trajectory. The overall safety is guaranteed as long as the initial state is ICS-free (which is something that can be reasonably assumed). Now, determining whether a given state of $\mathcal{A}$ is an ICS or not is a complex task since it requires to consider all possible future trajectories for $\mathcal{A}$. However, it is possible to take advantage of the approximation property demonstrated in [46] in order to compute a conservative approximation of the set of ICS. This is done by considering only a subset $\mathcal{I}$ of the full set of possible future trajectories.

3.4.3 Case Study

In this section we present the application of PMP to the case of a car-like vehicle $\mathcal{A}$ moving on a *planar* surface $\mathcal{W}$ and within a fully observable environment cluttered with stationary and dynamic obstacles. A control of $\mathcal{A}$ is defined by the couple (α, γ) where α is the rear wheel linear acceleration. and γ the steering velocity. The motion of $\mathcal{A}$ is governed by the following differential equations:

$$\begin{bmatrix} \dot{x} \\ \dot{y} \\ \dot{\theta} \\ \dot{v}_r \\ \dot{\xi} \end{bmatrix} = \begin{bmatrix} v_r \cos\theta \\ v_r \sin\theta \\ \frac{\tan \xi v_r}{L} \\ 0 \\ 0 \end{bmatrix} + \begin{bmatrix} 0 \\ 0 \\ 0 \\ 1 \\ 0 \end{bmatrix} \alpha + \begin{bmatrix} 0 \\ 0 \\ 0 \\ 0 \\ 1 \end{bmatrix} \gamma \qquad (3.32)$$

with $\alpha \in [\alpha_{min}, \alpha_{max}]$ (acceleration bounds), $\gamma \in [\gamma_{min}, \gamma_{max}]$ (steering velocity bounds), and $|\xi| \leq \xi_{max}$ (steering angle bounds). L is the wheelbase of $\mathcal{A}$.

For practical reasons, the duration of the trajectories of $\mathcal{I}$ has to be limited to a given *time horizon* that determines the overall level of safety of $\mathcal{A}$. In our case, the subset $\mathcal{I}$ considered in order to compute a conservative approximation of the set of ICS includes the braking trajectories with a constant control selected from $[(\alpha_{min}, \dot{\xi}_{max}), (\alpha_{min}, 0), (\alpha_{min}, \dot{\xi}_{min})]$, and applied over the time necessary for $\mathcal{A}$ to stop.

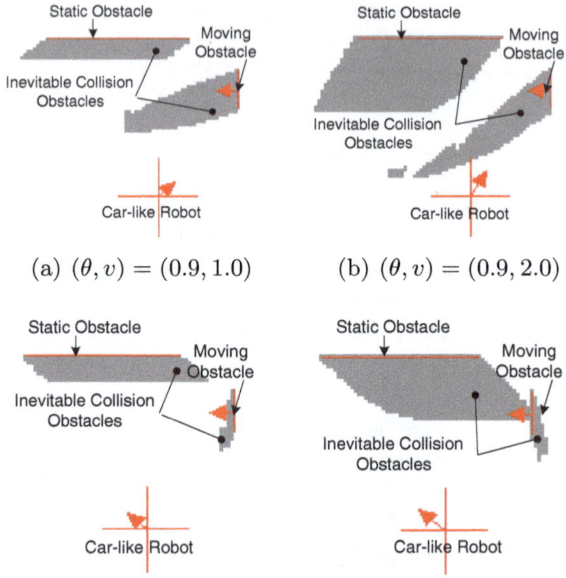

(a) $(\theta, v) = (0.9, 1.0)$ (b) $(\theta, v) = (0.9, 2.0)$

(c) $(\theta, v) = (2.2, 1.0)$ (d) $(\theta, v) = (2.2, 2.0)$

Fig. 3.24. (θ, v)-slices of the state space of $\mathcal{A}$. Shaded regions are ICS respectively defined for the braking trajectory of control $(\alpha_{min}, \dot{\xi}_{min})$ (top), and all braking trajectories with controls selected from $[(\alpha_{min}, \dot{\xi}_{max}), (\alpha_{min}, 0), (\alpha_{min}, \dot{\xi}_{min})]$ (bottom).

Fig. 3.24 depicts the ICS obtained when different set of braking trajectories are considered. Each subfigure represents a (θ, v)-slice of the full 5D state space of $\mathcal{A}$. In the top subfigures, only the braking trajectory of control $(\alpha_{min}, \dot{\xi}_{min})$ is considered. In the bottom subfigures, the three braking trajectories are considered.

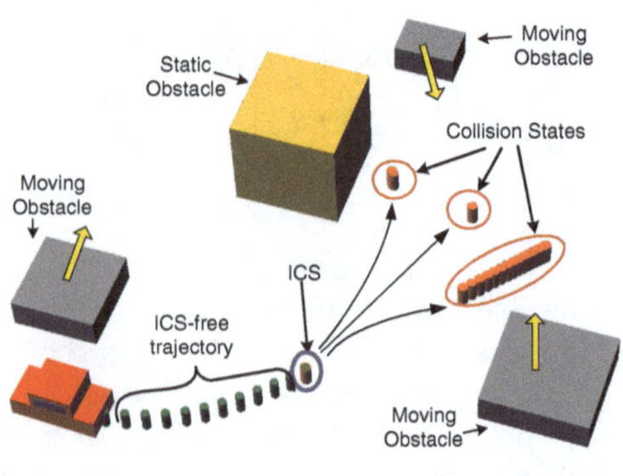

Fig. 3.25. The state labelled ICS is an ICS since the three braking trajectories issued from it yield collisions

In PMP, checking whether a state is an ICS or not is carried out by testing if all the braking trajectories yield a collision with one of the moving obstacles. If so, the state is an ICS. In fig. 3.25), the collision states in red represent the collision that will occur in the future from this state for all trajectories of $\mathcal{I}$. In this case, since all trajectories collide in the future, this state is an ICS. In PMP, every new state is similarly checked to be an ICS or not over $\mathcal{I}$. In case all the trajectories appear to be in collision in the future, this state is an ICS and is not selected.

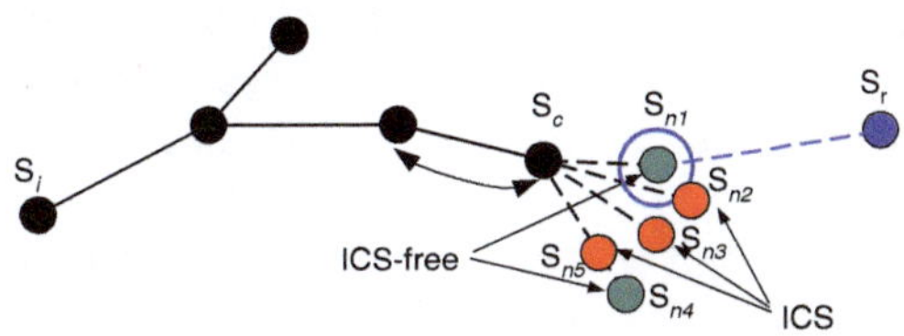

Fig. 3.26. Search tree construction principle

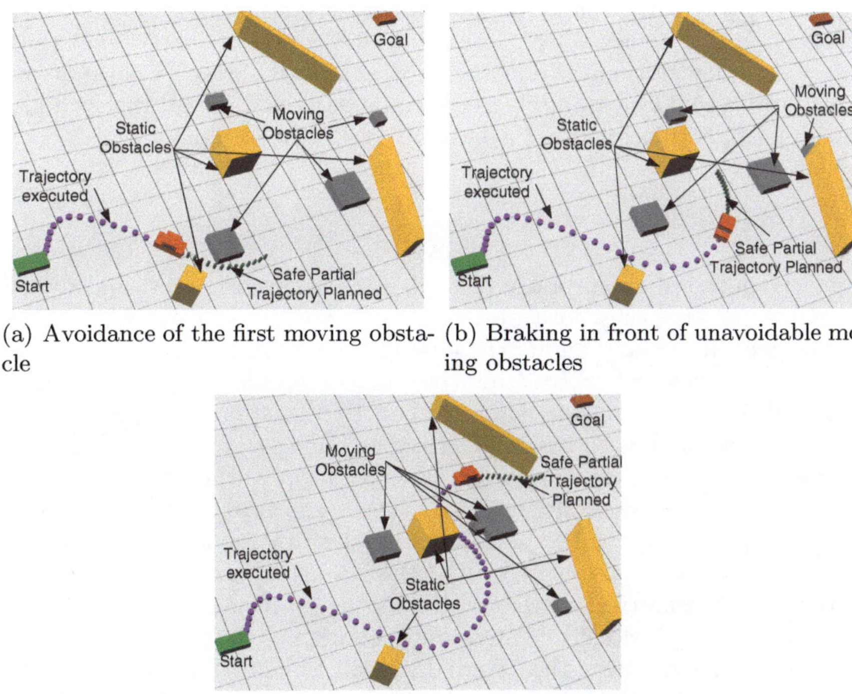

(a) Avoidance of the first moving obstacle

(b) Braking in front of unavoidable moving obstacles

(c) Avoidance of the last static obstacle

Fig. 3.27. Results of a 2D safe Partial Motion Planning ($\delta_c = 1s$, $v_{max} = 2.0m/s$, $\xi_{max} = \pi/3rad$, $\dot{\xi}_{max} = 0.2rad/s$, $\alpha_{max} = 0.1m/s^2$)

The exploration method used is the well known Rapidly-Exploring Random Tree method (RRT) [47]. RRT incrementally builds a tree in the state space of $\mathcal{A}$. The basic principle of RRT is depicted in Fig. 3.26. A state s_r is randomly selected first. Then, the closest node in the tree, say s_c, is determined. Constant controls selected from $\mathcal{U}_{2D}$=[$(\alpha_{max}, 0)$; $(\alpha_{max}, \dot{\xi}_{max})$; $(\alpha_{max}, \dot{\xi}_{min})$;$(0, \dot{\xi}_{max})$; $(0,0)$; $(0, \dot{\xi}_{min})$;$(\alpha_{min}, \dot{\xi}_{max})$; $(\alpha_{min}, 0)$; $(\alpha_{min}, \dot{\xi}_{min})$] are then applied to s_c for a duration ϵ, they yield a set of candidate trajectories ending in given states s_{ni}. These candidate trajectories are pruned out: only are kept the trajectories that are collision-free and whose final state is ICS-free (as per property 2, such trajectories are ICS-free). Finally, the trajectory whose final state is closer to s_r is selected and added up to the tree. This process is repeated until the end of the time available where the best partial trajectory extracted from the tree is returned.

In Fig. 3.27 we can see an example of a navigation from a still starting state (green box) to a still goal state (red box). The environment is cluttered with moving (grey) and static (orange) obstacles. In 27(a) one can observe how the safe partial trajectory (green) is calculated and planned within the time-state space in order to avoid the obstacle moving upward. The states in blue behind the car, is the trajectory, built from partial trajectories from the previous PMP cycles and (ideally) executed by the robot. In 27(b) we can observe that the car was obliged to slow down at the intersection of several obstacles, since no other safe trajectories could be found, before to re-accelerate. In 27(c) the system has planned a partial trajectory that avoids the last static obstacle.

3.5 Conclusion

This paper addressed the problem of navigating safely in a open and dynamic environment sensed using both on-board and external sensors. After a short presentation of the context and of the related open problems, we focused on two complementary questions: how to interpret and to predict the motions and the behaviors of the sensed moving entities ? how to take appropriate goal-oriented navigation decisions in such a rapidly changing and sensed environment?

In order to answer these questions, we have proposed an approach including three main complementary functions: (1) Scene interpretation and short-term motion prediction for the sensed potential obstacles, using the "Bayesian Occupancy Filtering" approach (*BOF*) (2) Medium-term motion and behavior prediction for the observed entities, using motion pattern learning and hierarchical Hidden Markov Models; (3) On-line goal-oriented navigation decision in a dynamic environment, using the "Partial Motion Planning" paradigm (*PMP*).

The first function (*BOF*)has experimentally been validated on our experimental vehicle (the Cycab), for avoiding partially observed moving obstacles. A scenario involving the Cycab, a moving pedestrian, and a parked car which temporarily hide the pedestrian to the sensors of the Cycab, has successfully been executed. In this experiment, the avoidance behavior has been obtained by combining the *occupancy probability* and the *danger probability* of each cell of the grid. The second function has experimentally been validated on some indoor data (the INRIA entry hall), using about 1000 tracked human trajectories for the initial learning phase. At the moment, the last function (PMP) has only been experimented in simulation.

Current work mainly deals with three major points: (1) Improvement of the prediction approaches for making it possible to cope with larger environments (such as complex urban traffic situations), while preserving the efficiency property; the current development on the WOG model is an example of this work. (2) Fusion of our current interpretation and prediction paradigms with higher-level information (e.g. GPS maps, moving entities properties, nominal behaviors ...) to better estimate the scene participants behaviors. (3) Integration of the three previous functions, and implementation and test this new navigation system on our experimental platform involving the INRIA car park, several Cycabs, and both inboard and infrastructure sensors.

Acknowledgements

This work has been partially supported by several national and international projects and funding bodies: European projects *Carsense, CyberCars & PreVent/ProFusion*; French Predit projects *Puvame & MobiVip*. The authors would like to thanks Pierre Bessiére, Christophe Couéé and Cedric Pradalier for their contributions and fruitful discussions.

References

1. C. Coué, C. Pradalier, and Laugier C. Bayesian programming for multi-target tracking: an automotive application. In *Proceedings of the International Conference on Field and Service Robotics*, Lake Yamanaka (JP), July 2003.
2. H.P. Moravec. Sensor fusion in certainty grids for mobile robots. *AI Magazine*, 9(2), 1988.
3. A. Elfes. Using occupancy grids for mobile robot perception and navigation. *IEEE Computer, Special Issue on Autonomous Intelligent Machines*, Juin 1989.
4. S. Thrun. Robotic mapping: A survey. In *Exploring Artificial Intelligence in the New Millenium*. Morgan Kaufmann, 2002.
5. E. Prassler, J. Scholz, and A. Elfes. Tracking multiple moving objects for real-time robot navigation. *Autonomous Robots*, 8(2), 2000.
6. O. Lebeltel. *Programmation Bayésienne des Robots*. Thèse de doctorat, Institut National Polytechnique de Grenoble, Grenoble, France, Septembre 1999.
7. O. Lebeltel, P. Bessière, J. Diard, and E. Mazer. Bayesian robot programming. *Autonomous Robots*, 16:49–79, 2004.
8. C. Coué and P. Bessière. Chasing an elusive target with a mobile robot. In *Proceedings of the IEEE-RSJ International Conference on Intelligent Robots and Systems*, Hawai (HI), 2001.

9. A. H. Jazwinsky. *Stochastic Processes and Filtering Theory.* New York : Academic Press, 1970.
10. R.E. Kalman. A new approach to linear filtering and prediction problems. *Journal of basic Engineering*, 35, Mars 1960.
11. G. Welch and G. Bishop. An introduction to the Kalman filter. available at http://www.cs.unc.edu/ welch/kalman/index.html.
12. I. Daubechies. *Ten Lectures on Wavelets.* Number 61 in CBMS-NSF Series in Applied Mathematics. SIAM Publications, Philadelphia, 1992.
13. Stéphane Mallat. *A Wavelet Tour of Signal Processing.* Academic Press, San Diego, 1998.
14. D. K. Pai and L.-M. Reissell. Multiresolution rough terrain motion planning. In *IEEE Transactions on Robotics and Automation*, volume 1, pages 19–33, February 1998.
15. Bruno Sinopoli, Mario Micheli, Gianluca Donato, and T. John Koo. Vision based navigation for an unmanned aerial vehicle. In *Proceedings of the International Conference on Robotics and Automation*, May 2001.
16. Manuel Yguel, Olivier Aycard, and Christian Laugier. Internal report: Wavelet occupancy grids: a method for compact map building. Technical report, INRIA, 2005.
17. C. Pradalier and S. Sekhavat. Simultaneous localization and mapping using the geometric projection filter and correspondence graph matching. *Advanced Robotics*, 2004. To appear.
18. J.H. Reif and M. Sharir. Motion planning in the presence of moving obstacles. In *Symp. on the Foundations of Computer Science*, pages 144–154, Portland (OR) USA, October 1985.
19. Qiuming Zhu. A stochastic algorithm for obstacle motion prediction in visual guidance of robot motion. *The IEEE International Conference on Systems Engineering*, pages 216–219, 1990.
20. R.E. Kalman. A new approach to linear filtering and prediction problems. *Trans. Am. Soc. Mech. Eng., Series D, Journal of Basic Engineering*, 82:35–45, 1960.
21. Dizan Vasquez and Thierry Fraichard. Motion prediction for moving objects: a statistical approach. In *Proc. of the IEEE Int. Conf. on Robotics and Automation*, pages 3931–3936, New Orleans, LA (US), apr 2004.
22. Sarah Osentoski, Victoria Manfredi, and Sridhar Mahadevan. Learning hierarchical models of activity. In *IEEE/RSJ International Conference on Intelligent Robots and Systems*, Sendai, Japan, 2004.
23. Maren Bennewitz, Wolfram Burgard, Grzegorz Cielniak, and Sebastian Thrun. Learning motion patterns of people for compliant robot motion. *Internationl Journal of Robotics Research*, 24(1):31–48, January 2005.
24. Lawrence R. Rabiner. A tutorial on hidden markov models and selected applications in speech recognition. *Readings in speech recognition*, pages 267–296, 1990.
25. Bernd Fritzke. A growing neural gas network learns topologies. *Advances in Neural Information Processing Systems*, 1995.
26. Kevin Patrick Murphy. *Dynamic Bayesian Networks: Representation, Inference and Learning.* PhD thesis, University of California, Berkeley (USA), 2002.
27. Matthew Brand. Structure learning in conditional probability models via an entropic prior and parameter extinction. Technical report, MERL a Mitsubishi Electric Research Laboratory, 1998.

28. Henri Binsztok and Thierry Artières. Learning model structure from data: an application to on-line handwriting. *Electronic Letter on Computer Vision and Image Analyisis*, 5(2), 2005.
29. Andrew J. Viterbi. Error bounds for convolutional codes and an asymptotically optimum decoding algorithm. *IEEE Transactions on Information Theory*, IT-13(2):260–269, April 1967.
30. N. Dempster, A.and Laird, , and D. Rubin. Maximum likelihood from incomplete data via the EM algorithm. *Journal of the Royal Statistical Society*, 9(1):1–38, 1977. Series B.
31. Nir Friedman. Learning belief networks in the presence of missing values and hidden variables. In *Proc. of the Fourteenth International Conference on Machine Learning*, pages 125–133, 1997.
32. Andreas Stolcke and Stephen Omohundro. Hidden markov model induction by bayesian model merging. In Stephen José Hanson, Jack D. Cowan, and C. Lee Giles, editors, *Advances in Neural Information Processing Systems*, volume 5, pages 11–18, Denver, USA, 1993. Morgan Kaufmann, San Mateo, CA.
33. A. Caporossi, D. Hall, P. Reignier, and J.L. Crowley. Robust visual tracking from dynamic control of processing. In *International Workshop on Performance Evaluation of Tracking and Surveillance*, pages 23–31, Prague, Czech Republic, May 2004.
34. J. H. Reif and M. Sharir. Motion planning in the presence of moving obstacles. In *Proc. of the IEEE Symp. on the Foundations of Computer Science*, pages 144–154, Portland, OR (US), October 1985.
35. R.C. Arkin. Motor schema-based mobile robot navigation. *International Journal of Robotics Research*, 8(4):92–112, August 1989.
36. J. Borenstein and Y. Koren. The vector field histogram - fast obstacle avoidance for mobile robots. *IEEE Journal of Robotics and Automation*, 7(3):278–288, June 1991.
37. R. Simmons and S. Koenig. Probabilistic robot navigation in partially observable environments. In *Proceedings of the International Joint Conference on Artificial Intelligence*, pages 1080–1087, 1995.
38. D. Fox, W. Burgard, and S. Thrun. The dynamic window approach to collision avoidance. Technical Report IAI-TR-95-13, 1 1995.
39. M. Khatib. *Sensor-based motion control for mobile robots.* PhD thesis, LAAS-CNRS December, 1996, 1996.
40. R. Simmons. The curvature velocity method for local obstacle avoidance. In *Proceedings of the International Conference on Robotics and Automation*, pages 3375–3382, Minneapolis (USA), april 1996.
41. P. Fiorini and Z. Shiller. Motion planning in dynamic environments using velocity obstacles. *International Journal of Robotics Research*, 17(7):760–772, July 1998.
42. N.Y. Ko and R. Simmons. The lane-curvature method for local obstacle avoidance. In *Proceedings of the IEEE/RSJ International Conference on Intelligent Robots and Systems*, Victoria (Canada), October 1998.
43. J. Minguez and L. Montano. Nearness diagram navigation (ND): A new real time collision avoidance approach for holonomic and no holonomic mobile robots. In *Proceedings of the IEEE/RSJ International Conference on Intelligent Robots and Systems*, Takamatsu, Japan, November 2000.
44. N. Roy and S. Thrun. Motion planning through policy search. In *Proceedings of the IEEE/RSJ International Conference on Intelligent Robots and Systems*, Lausanne, Switzerland, 2002.

45. J. Minguez, L. Montano, and J. Santos-Victor. Reactive navigation for non-holonomic robots using the ego kinematic space. In *Proceedings IEEE International Conference on Robotics and Automation*, Washington (US), May 2002.
46. Thierry Fraichard and Hajime Asama. Inevitable collision states - a step towards safer robots? *Advanced Robotics*, 18(10):1001–1024, 2004.
47. S. LaValle and J. Kuffner. Randomized kinodynamic planning. In *Proceedings of the IEEE International Conference on Robotics and Automation*, volume 1, pages 473–479, Detroit (US), May 1999.

Part II
Obstacle Avoidance and Motion Planning in Dynamic Environments

Provably Safe Motions Strategies for Mobile Robots in Dynamic Domains

Rachid Alami, K. Madhava Krishna, and Thierry Siméon

LAAS-CNRS
7, Avenue du Colonel Roche - 31077 Toulouse - France

Summary. We present in this paper a methodology for computing the maximum velocity profile over a trajectory planned for a mobile robot. Environment and robot dynamics as well as the constraints of the robot sensors determine the profile. The planned profile is indicative of maximum speeds that can be possessed by the robot along its path without colliding with any of the mobile objects that could intercept its future trajectory. The mobile objects could be arbitrary in number and the only information available is their maximum possible velocity. The velocity profile also enables to deform planned trajectories for better trajectory time. The methodology has been adopted for holonomic and non-holonomic motion planners. An extension of the approach to an online real-time scheme that modifies and adapts the path as well as velocities to changes in the environment such that both safety and execution time are not compromised is also presented for the holonomic case. Simulation and experimental results illustrate the efficacy of this methodology.

4.1 Introduction

Several strategies exist for planning collision free paths in an environment whose model is known [9]. However during execution, parameters such as robot and environment dynamics, sensory capacities need to be incorporated for safe navigation. This is especially so if the robot navigates in an area where there are other mobile objects such as humans. For example in Figure 4.1, the robot would require to slow down as it approaches the doorway, in anticipation of mobile objects to emerge from there, even if it does not intend to make a turn through the doorway. A possible means to tackle the above problem at the execution stage is to always navigate the robot at very small speeds. In fact, reactive schemes such as the nearness diagram approach [11] operate the robot at minimal velocities throughout the navigation. However incorporating the computation of a velocity profile at the planning stage would circumvent not only the problem of permanent conservative velocities throughout navigation but would also allow for a modification of the trajectory to achieve lower time such as shown in Figure 4.2.

A novel pro-active strategy that incorporates robot and environment dynamics as well as sensory constraints onto a collision free motion plan is presented in this paper. By pro-active we mean that the robot is always in a state of expectation regarding the possibility of a mobile object impinging onto its path from regions

C. Laugier and R. Chatila (Eds.): Autonomous Navi. in Dyn. Environ., STAR 35, pp. 85–106, 2007.
springerlink.com

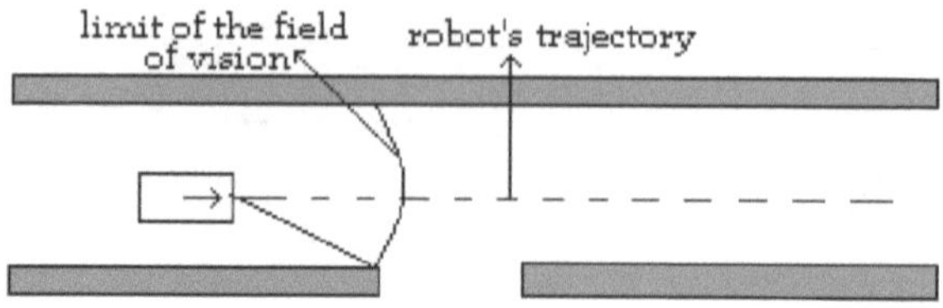

Fig. 4.1. A safe robot has to slow down while approaching the doorway

invisible to its sensor. This pro-active state is reflected in the velocity profile of the robot, which guarantees that in the worst case scenario, the robot will not collide with any of the moving objects that can cross its path. The ability of the algorithm to compute a priori velocities for the entire trajectory accounting for moving objects moving in arbitrary directions is the essential novelty of this effort. A similar kind of strategy for the aforementioned objective does not appear to have been presented in the robotic literature so far.

As is always the case, planned paths and profiles need constant modification at the execution stage due to changes in the environment. For example, a profile and path that was planned for an environment with a closed doorway needs to be modified during real-time if the doorway is found open. The paper also addresses the problem illustrated in Figure 4.3. Given an initial trajectory planned for a particular environment, how does the robot modify its trajectory while new objects (not necessarily intersecting the robot's trajectory) are introduced into the environment such that the basic philosophy of ensuring safety as well as reducing time lengths of the path continue to be respected. Simulation and experimental results are presented to indicate the efficacy of the scheme. In [1] we had reported how the maximum velocity profiles can be computed for any generic planner and in [8] we presented initial simulation and experimental results of the reactive version of [1].

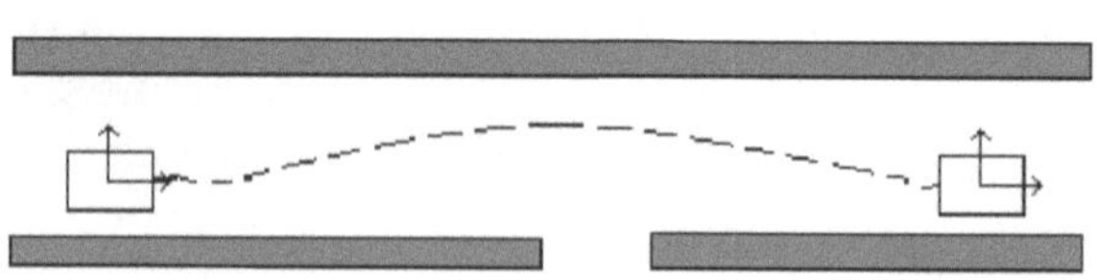

Fig. 4.2. A longer path could be faster due to higher velocity

Related work can be cited in the areas of modifying global plans using sensory data obtained during execution for overcoming uncertainty accumulated during motions [3] and those that try to bridge the gap between planning and uncertainty [10] or planning and control [7], [2]. The velocity obstacle concept[13][5] bears resemblance to the current endeavor in that they involve selection of a robot velocity that avoids any number of moving objects. The difference is that in the present approach the only information about the mobile object available is

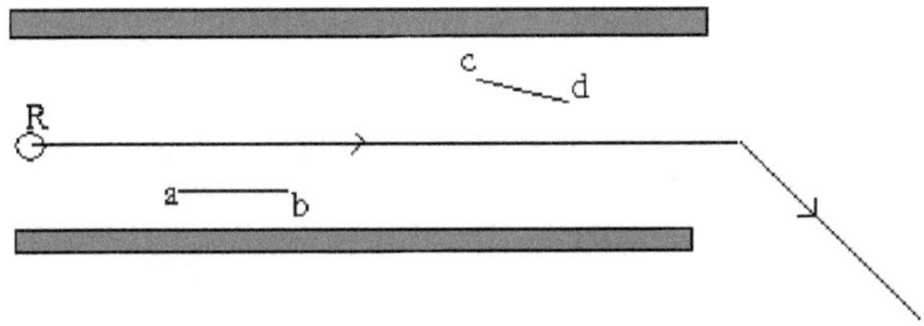

Fig. 4.3. How does the robot adapt its path in the presence of new segments (a, b) and (c, d) while maintaining safety

the bound on velocity. The direction of motion and the actual velocities are not known during computation of the velocity profile. The work of Strachniss [14] also involves considering the robot's pose and velocities at the planning phase. A path is determined in the (x, y) space and a subgoal is chosen. A sequence of linear and angular velocities, (v, w), is furnished till the subgoal is reached. It however does not speak of reducing the time of the path by modifying it, and the dynamics of the environment does not seem to affect the computation of the velocity profile. In [12] a policy search approach is presented that projects a low dimensional intermediate plan to a higher dimensional space where the orientation and velocity are included. As a result better motion plans are generated that enable better execution of the plan by the robot. The current effort has similar ties to [12] at the planning level but also extends it to a suitable reactive level in the presence of new obstacles encountered during execution.

4.2 Problem Definition

Consider a workspace cluttered by static polygonal obstacles $\mathcal{O}_i$. The static obstacles can hide possible mobile objects whose motions are not predictable; the only information is their bounded velocity v_{ob}. The robot $\mathcal{R}$, modelled as a disc, is equipped with an omnidirectional sensor having a limited range R_{vis}. We call $\mathcal{C}_{vis}$ the visibility circle, centered at robot's position with radius R_{vis}. The paths of $\mathcal{R}$ are sequences of straight segments possibly connected with circular arcs of radius ρ in case of a non-holonomic robot. The robot's motion is subject to dynamic constraints simply modelled by a bounded linear velocity $v \in [0, v_{rm}]$ and a bounded acceleration $a \in [-a_{-m}, a_m]$. The maximum possible deceleration a_{-m} needs not equal the maximum acceleration a_m. The following problems are addressed in the paper:

Problem 1: Given a robot's path $\tau(s)$ computed by a standard planner (e.g. [9]), determine the maximal velocity profile $v_\tau(s)$ such that, considering the constraints imposed by its dynamics, the robot can stop before collision occurs with any of the mobile objects that could emerge from regions not visible to the robot at position $s \in \tau(s)$. For example the velocity profile dictates that the robot in Figure 4.1 slows down near the doorway in expectation of mobile objects from the other side.

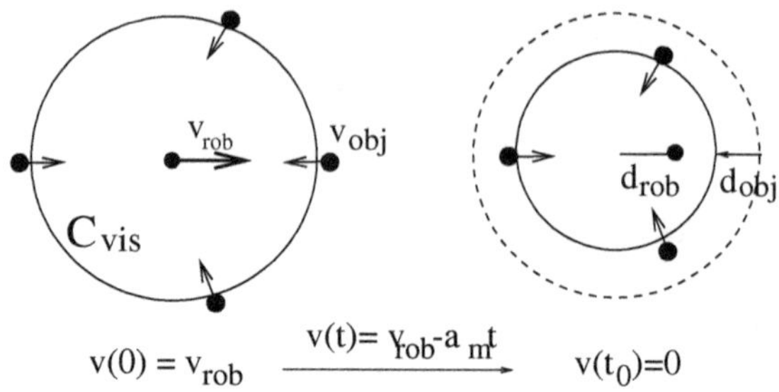

Fig. 4.4. Mobile objects may appear anywhere on C_{vis}'s contour

We call $MP = (\tau(s), v_\tau(s))$ a **robust motion plan**. The velocity profile allows us to define the time $T(\tau)$ required for the robust execution of path τ:

$$T(\tau) = \int_0^L \frac{ds}{v_\tau(s)}.$$

Problem 2: Modify the planned trajectory such that the overall trajectory time $T(\tau)$ is reduced. For example, the path of Figure 4.2 is traversed in a shorter time, though longer path, than the one of Figure 4.1.

Problem 3: Adapt the path and velocities reactively in the presence of new objects not a part of the original workspace such that the criteria of safe velocities and reduced time of path continue to be respected. This is illustrated in Figure 4.3.

4.3 From Path to Robust Motion Plan

The procedure for computing the maximum velocity profile $v_\tau(s)$ delineated in Sections 4.3.1, 4.3.2 and 4.3.3 addresses the first problem. The constraints imposed by the environment on the robot's velocity are due to two categories of mobile objects. The first category consists of mobiles that could appear from anywhere outside the boundary of the visibility circle C_{vis}. The second category involves mobile objects that could emerge from shadows created in C_{vis} due to stationary objects.

4.3.1 Velocity Constraints Due to the Environment

No obstacles in C_{vis}: In the simple case where the robot's position is such that no static obstacle lies inside C_{vis}, a moving object may appear (at time $t = 0$) anywhere on C_{vis}'s boundary (Fig. 4.4). Let V_{rb} denote the maximum possible robot velocity due to a mobile object at the boundary. At time $t_0 = v_{rb}/a_{-m}$ (i.e., when the robot is stopped), the distance crossed by the object is $d_{obj}(v_{rb}) \leq v_{ob}v_{rb}/a_{-m}$. Avoiding any potential collision imposes that $R_{vis} \geq$

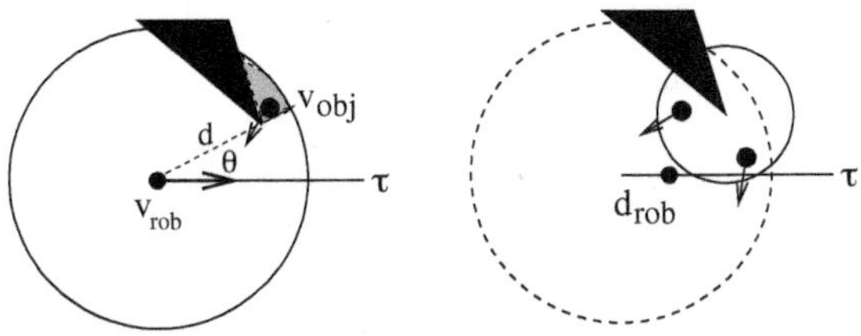

Fig. 4.5. Mobile objects may also appear from the shadows of static obstacles

$d_{rb}(v_{rb}) + d_{obj}(v_{rb})$, where $d_{rb} = v_{rb}^2/2a_{-m}$. The condition relates v_{rb} to the sensor's range R_{vis} as:

$$v_{rb} = -v_{ob} + \sqrt{v_{ob}^2 + 2a_{-m}R_{vis}} \tag{4.1}$$

Influence of shadowing corners: Static obstacles lying inside $\mathcal{C}_{vis}$ may create shadows (e.g., see the grey region of Figure 4.5) which contains mobile objects. The worst-case situation occurs when the mobile object remains unseen until it arrives at the *shadowing corner* of a polygonal obstacle. Since the mobile object's motion direction is not known it is best modeled for a worst case scenario as an expanding circular wave of radius $v_{ob}t$ centered at (d, θ). Its equation is given by: $(X(t) - d\cos\theta)^2 + (Y(t) - d\sin\theta)^2 = v_{ob}^2 t^2$. Let us first consider that the robot's path τ is a straight segment. Considering that the intersections between the circular wave and the robot's segment path, should never reach the robot before it stops at time $t_0 = v_{rs}/a_{-m}$ yields the following velocity constraint:

$$v_{rsv}^4 - 4(a_{-m}d\cos\theta + v_{ob}^2)v_{rsv}^2 + 4a_{-m}^2 d^2 \geq 0 \tag{4.2}$$

Here v_{rsv} is the maximum possible robot velocity due to the shadowing vertex under consideration. The solution of Equation 4.2 gives v_{rsv}, as a function of (d, θ). This solution only exists under the condition $v_{ob} > \sqrt{a_{-m}d(1 - \cos\theta)}$, i.e., when the object's velocity v_{ob} is sufficiently high to interfere with the robot's halting path. Otherwise, the shadowing corner does not constrain the robot's velocity which can be set to the maximum bound v_{rm}.

Similar reasoning can be applied to the case where the robot traverses a circular arc path of radius ρ. This case however leads to a nonlinear equation that needs to be solved numerically to derive the maximal velocity [4]. The expression that needs to be solved for computing the maximum velocity at a given point on a circular arc is of the form:

$$((v_{rsv}^2 v_{ob}^2)/a_{-m}^2) + 2\rho^2\cos(v_{ob}^2/2a_{-m}\rho) + 2d\rho\sin((v_{ob}^2/2a_{-m}\rho) - \theta) = d^2 + 2\rho^2 - 2d\rho\sin\theta \tag{4.3}$$

4.3.2 Computing the Shadowing Corners

The problem of determining the set of shadowing corners needed for the velocity computation in Section 4.3.1 is the problem of extracting those vertices of the polygonal obstacle to which a ray emitted from the robot's center is tangential (Fig. 4.6). The set of shadowing corners can be easily extracted from an algorithm that outputs the visibility polygon [15] as a sorted list of vertices.

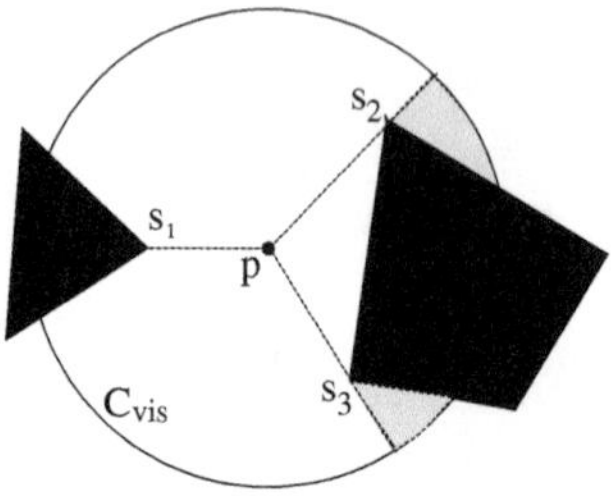

Fig. 4.6. Shadowing corners: among the three vertices of $\mathcal{V}(p)$, only s_2 and s_3 create shadows (the line going through s_1 is not tangent to the left obstacle)

4.3.3 Computation of Maximum Velocity Profile $v_\tau(s)$

While the methodology for computing the maximum velocity profile delineated here is essentially for a holonomic path, its extension to the non-holonomic case is not difficult.

1. A holonomic path τ, consisting of a sequence of straight line segments ab, bc, cd (Fig. 4.7) is deformed into a sequence of straight lines and clothoids to ensure continuity of velocities at the bends [6]. The maximum deviation from an endpoint to its clothoidal arc (depicted as e in Figure 4.7) is dependent on the nearest distance to an object from the endpoint under consideration.
2. The linear velocity along a clothoid is a constant and the maximum possible linear velocity considering robot dynamics alone is calculated for each of the clothoidal arc $a1b0, b1c0$ according to [6] and is represented as $v_c(a1), v_c(b1)$.
3. The straight segment $aa1$ is discretized into M equally spaced points, excluding the endpoints of the segment, viz a and $a1$. We denote the first such point as a_1 and the last such point as a_M. The point of entry into the clothoid, viz. $a1$ is also denoted as a_{M+1}.
4. For each of the N points, a_i, the steps $4a$ to $4e$ are repeated.

4a Maximum possible velocity that a robot could have such that it can come to a halt before colliding with objects that enter into the robot's field of vision from the boundary is computed as $v_{rb}(a_i)$ according to Eq. 4.1

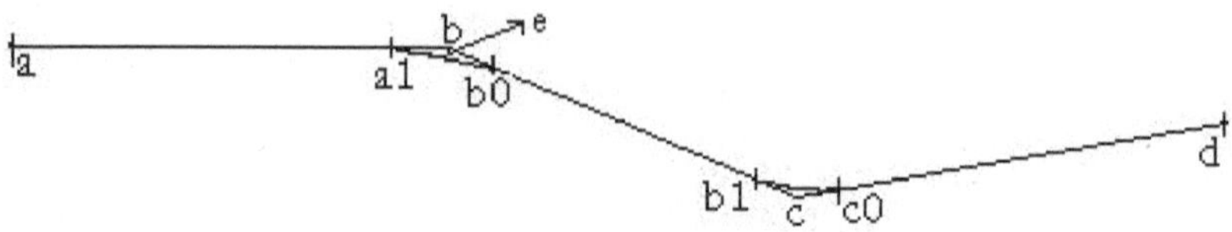

Fig. 4.7. A holonomic path deformed into a sequence of straight segments and clothoidal arcs

4b Velocity of the robot due to stationary obstacles inside the robot's field of vision that create shadows is computed as $v_{rsv}(a_i)$ according to Eq. 4.2. The minimum of all the velocities due to such vertices is found and denoted as $v_{rs}(a_i)$.

4c The maximum possible velocity of the robot at a_i due to environment is then computed as:

$$v_{re}(a_i) = min(v_{rb}(a_i), v_{rs}(a_i)) \tag{4.4}$$

4d Velocity of the robot at a_i due to its own dynamics is given by:

$$v_{rd}(a_i) = \sqrt{v_r^2(a_{i-1}) + 2a_m s(a_i, a_{i-1})} \tag{4.5}$$

The above equation is computed if $v_{re}(a_i) > v_r(a_{i-1})$. Here $s(a_i, a_{i-1})$ represents the distance between the points a_i and a_{i-1} and a_m represents the maximum acceleration of the robot.

4e The eventual velocity at a_i is given by:

$$v_r(a_i) = min(v_{rd}(a_i), v_{re}(a_i), v_{rm}) \tag{4.6}$$

Here v_{rm} represents the maximum robot velocity permissible due to servo motor constants.

5. The velocity at the endpoint $a1$ is computed as $v_r(a1) = min(v_r(a1), v_c(a1))$ and this would be the linear velocity with which the robot would traverse the clothoid.
6. Steps $6a$ and $6b$ are performed by going backwards on each of the N points from a_N to a_1

6a If $v_r(a_i) > v_r(a_{i+1})$ then the modified maximum possible velocity at a_i is computed as:

$$v_{rd}(a_i) = \sqrt{v_r^2(a_{i+1}) + 2a_{-m} s(a_i, a_{i+1})} \tag{4.7}$$

6b Finally the maximum safe velocity at a_i is given as $v_r(a_i) = min(v_r(a_i), v_{rd}(a_i))$.

7. Repeat steps 3 to 6 for all the remaining straight segments to obtain the maximal velocity profile over a given trajectory τ as $v_\tau(s) = \{v_r(a), v_r(a_1), ..., v_r(a_N), v_r(a1), v_r(b_1), ..., v_r(d)\}$.

4.3.4 Modifying Planned Trajectory for Lower Time

The knowledge of the maximum velocity profile over a trajectory is utilized to tackle the problem posed in Section 4.2 of reducing the overall trajectory time of the path. The procedure for reducing trajectory time at the planning stage involves random deformation of the planned path and evaluating time along this path. The modified path becomes the new trajectory if time along it is less than along the original trajectory. The process is continued till over a finite number of attempts no further minimization of trajectory time is possible. Prior to delineating the algorithm it is to be noted that the set of all collision free space of the workspace is denoted as C_{free} and the current trajectory of the robot as $\tau_c(s)$. A point of discretization on a trajectory discretized into N parts is denoted

Algorithm 1. Globally reducing trajectory time

1: $N_{try} \leftarrow 0$
2: **while** $N_{try} < N_{attempts}$ **do**
3: Discretize current trajectory $\tau_c(s)$ into N_p where N_p is selected based on minimum discretization distance between two points.
4: Set $flag \leftarrow 0$
5: **for** $i = 1$ to N_p **do**
6: Compute minimum velocity at s_i due to shadowing vertices as $v_{rmin}(s_i)$
7: **if** $v_{rmin}(s_i) < v_{rm}$ **then**
8: Find a configuration $q(s_p) \in C_{free}$ and $s_p \notin \tau_c(s_k), k \in \{1, ..., N_p\}$ such that $q(s_p)$ is reachable from $q(s_i)$.
9: Find a point s_r on the remaining part of the trajectory, $s_r \in \tau_c(s_j); i < j \leq N_p$ such that $q(s_r)$ is reachable from $q(s_p)$.
10: Form a new trajectory through s_i, s_p, s_q and denote it as $\tau_n(s)$
11: **if** $T(\tau_n) < T(\tau_c)$ **then**
12: discretize τ_n into N_q points.
13: $\tau_c \leftarrow \tau_n$
14: $N_p \leftarrow N_q$
15: Set $flag \leftarrow 1$
16: **end if**
17: **end if**
18: **end for**
19: **if** $flag = 0$ **then**
20: $N_{try} \leftarrow N_{try} + 1$
21: **end if**
22: **end while**

as $p(s_i), i \in \{1, 2, ..., N\}$. The corresponding configuration of the robot at those points is denoted by $q(s_i)$.

The algorithm is given as Algorithm 1. Step 8 of the algorithm is carried out by searching for a collision free configuration which would displace the path away from the shadowing vertex responsible for the lowest velocity at s_i. Step 11 adapts the displaced path as the new current path if its trajectory time is less than the current path. $N_{attempts}$, is the number of unsuccessful attempts at minimizing trajectory time before the algorithm halts.

4.3.5 Remembering Sensor Information

The computation of the velocity profile at a given point on the robot's trajectory incorporates the robot's field of vision at that point. This field can change appreciably between two successive instances of computation. For example in Figure 4.8 the robot at position a has full field of vision of the corridor that is transverse to the robot's trajectory. However at position b the robot is blind to the zone shown in darker shade of gray. Hence it needs to slow down as it moves further down to c since it envisages the possibility of a moving object approaching it from the corners of the stationary objects. These corners are the starting areas of the robot's blind zone at b.

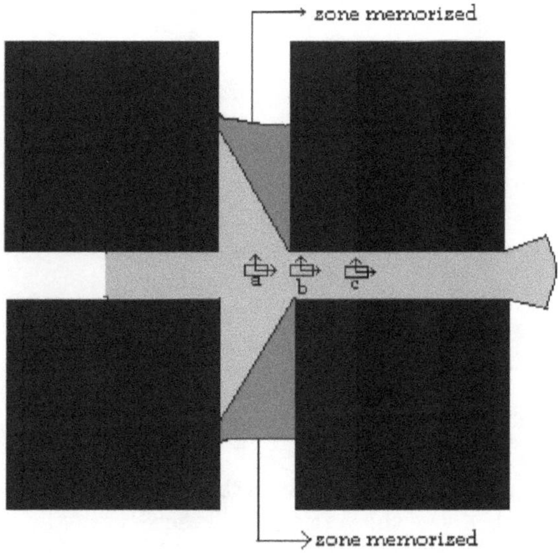

Fig. 4.8. Remembering of previous scenes

However, if the robot could remember the earlier scene, it could use this to compute its velocity profile during execution of the planned path. In such a case, if the robot did not see any moving objects in close proximity at a it can make use of this information at b to have a velocity profile from b that is greater than the one computed in the absence of such information. Figure 4.8 shows (in darker shade) the zone memorized by the robot. The contour of the memorized area represents the blind zone of the robot at b, from where mobile objects can emanate. The area in lighter shade of gray is the visibility polygon for the robot at b. With the passage of time the frontier of the remembered area shrinks due to the advancement of the imagined mobile objects from the initial frontier. The details of this scheme are given below.

Remembering is fruitful when a (hitherto non-shadowing) vertex begins to cast a shadow thereby hiding regions which were previously visible. The set of all vertices that are currently visible, shadowing and were at some prior instant visible, non-shadowing is denoted by $Vsns$. For every vertex $ve \in Vsns$ a corresponding vertex is associated and called the blind vertex. The blind vertices are of three categories explained in Figure 4.9 where the vertex a, non-shadowing for the robot at p becomes shadowing when the robot is at q. Correspondingly the vertex c of the triangular obstacle which was visible and shadowing when the robot was at p becomes invisible when the robot moves to q. Simultaneously one of the other end-points of b, viz. a, would also become inevitably invisible at q. Vertices like b fall in the second category. If b was already outside C_{vis} at p the intersection of C_{vis} with the segment ab, namely o is identified as the third category of blind vertex. The set of all such vertices is denoted by Vbs. These vertices are advanced by a distance $v_{ob}\Delta t$ where Δt is the time taken by the robot between p and q to new

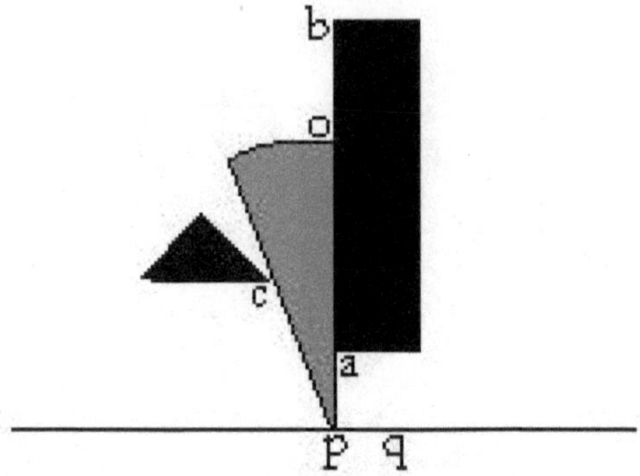

Fig. 4.9. Three categories of blind vertices

virtual locations along the line that connects those vertices to a. At q the velocity is computed due to the closest of the vertices in the set Vbs at their virtual locations instead of a, which is otherwise the vertex for which equation (4.2) is computed.. Such a trend continues till the distance between the robot to the closest hypothetical vertex is less than the actual distance of the robot to a. The remembering part of the algorithm is given in Algorithm 2. The set of all visible shadowing vertices is denoted by Vsh.

Algorithm 2. Remembering effects on velocity

for each vertex $v_e \in Vsh$ **do**
  if $v_e \in Vsns$ **then**
    for each vertex $v_b \in Vbs$ associated with v_e **do**
      Advance v_b by $v_{ob}\Delta t$
    end for
    Denote the distance from the robot's current location, s_c, to the closest of all advanced vertices, v_{bc} as d_{cvb}
    if $d(s_c, v_e) < d_{cvb}$ **then**
      Compute velocity due to the virtual vertex v_{bc} by Equation 4.2
    else
      Compute velocity due to the actual vertex v_e by Equation 4.2
    end if
  end if
end for

4.4 From Plan to Execution

The velocity profile, $v_\tau(s)$, is a sequence of maximum velocities calculated at discretized locations along the trajectory $\tau(s)$. The locations at which the velocity profile at the execution stage is computed are not the same locations as where the profile was computed during planning due to odometric and motor constraints.

Moreover, if there are changes in the environment it entails modifying the trajectory and hence the velocities. During execution it is computationally expensive to compute the profile for the entire remaining trajectory, hence the profile is computed for the next finite distance, given by, $d_{safe} = d_{max} + nd_{samp}$, where $d_{max} = v_{rm}^2/(2 * a_{-m})$, represents the distance required by the robot to come to a halt while it moves with the maximum permissible velocity afforded by motor constants. And $d_{samp} = v_{rm}t_{samp}$ is the maximum possible distance that the robot can move between two successive samples (time instants) of transmitting motion commands, where time between two samples is t_{samp}.

The main issue here is what should be the distance over which the velocity profile needs to be computed during execution such that it is safe. A velocity command is **not** considered safe if it is less than the current velocity and **not** attainable within the next sample. The velocity is constrained by the environment as well as robot's own dynamics and hence their role are both studied below.

Effect of Environment: Mobile objects that can emerge from corners in a head-on direction cause the greatest change in velocity over two samples. Figure 4.10 shows one such situation, where the rectangular object casts a shadow and is susceptible to hide mobile objects. Let the current velocity of the robot at a due to the object be v_1. Let the velocity at a distance, s, from a, at b (Fig. 4.10) due to the object be v_2. Velocities at a and b are given by:

$$v_a = -v_{ob} + \sqrt{v_{ob}^2 + 2a_{-m}d} \tag{4.8}$$

$$v_b = -v_{ob} + \sqrt{v_{ob}^2 + 2a_{-m}(d-s)} \tag{4.9}$$

Hence we have:

$$v_a^2 - v_b^2 = 2a_{-m}s + 2v_{ob}(\sqrt{v_{ob}^2 + 2a_{-m}(d-s)} - \sqrt{v_{ob}^2 + 2a_{-m}d}) \tag{4.10}$$

Evidently the second term on the right hand side of Equation 4.10 is negative, since the second square root term is less positive than the first. Hence $v_a^2 - v_b^2 \leq 2a_{-m}s$. Therefore the velocity at b, v_b can be attained from the velocity at a, v_a under maximum deceleration, d_m, irrespective of the maximum velocity of the mobile object or the robot's own motor constraints. This was intuitively expected since the robot's velocity at any location is the maximum possible velocity that guarantees immobility before collision; its velocity at a subsequent location permitted by the environment would be greater than or equal to the velocity at the same location obtained under maximum deceleration from the previous location. In other words, for safeness of velocity going purely by environmental considerations it suffices to calculate the velocity, for the next sampling distance alone, for without loss of generality, $d = d_{samp}$.

Effect of robot's dynamics: The robot needs to respect the velocity constraints imposed while nearing the clothoidal arcs and eventually while coming to the target. The robot can reach zero velocity from its maximum velocity over a distance of d_{max}, computed before. Hence $d_{max} + d_{samp}$ represents the safe distance over which the velocities need to be computed.

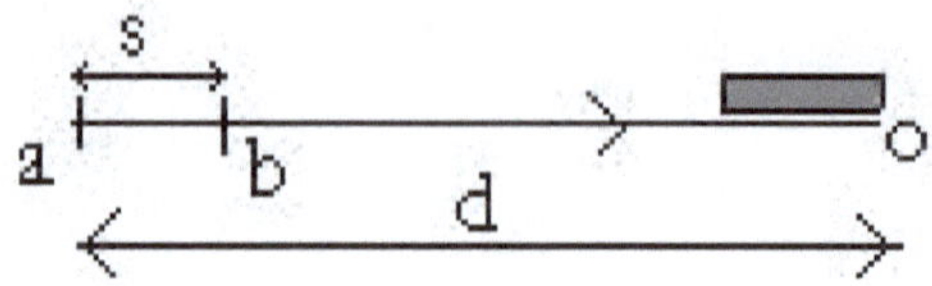

Fig. 4.10. The effect of the rectangular object that could hide possible mobile objects on the robot's velocity at locations a and b

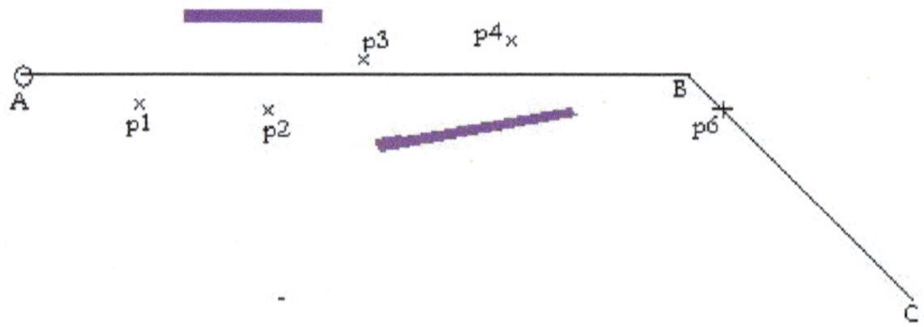

Fig. 4.11. A trajectory in the presence of new objects. The points marked with crosses represent locations through which a path is searched for reduced time trajectory.

4.4.1 Online Path Adaptation for Lower Trajectory Time

The third of the problems outlined in Section 4.2 is addressed here. During navigation the robot in general comes across objects hitherto not a part of the map. The robot reacts to these new objects in line with the basic philosophy of safety as well as time reduced paths. The adaptation proceeds by finding locations over a finite portion of the future trajectory where drops in velocity occur and pushing the trajectory away from those vertices of the objects that caused these drops to areas in free space where higher velocities are possible. A search is made through the newly found locations of higher velocities for a time reduced path.

Generalized Procedure: The generalized procedure for adapting the path in presence of new objects is delineated through Figure 4.11.

1. On the trajectory segment that is currently traversed, AB in Figure 4.11, enumerate the vertices of objects that reduce the velocity of the robot.
2. The positions are found on AB where the influence of vertices is likely to be maximal.
3. These positions are pushed by distances $d_p = k(v_l - v_r)$, where v_l and v_r are the velocities at that location on the path due to the most influential vertices on the left and right of the path. These new locations are denoted as $p1, p2, p3, p4$ (Fig. 4.11) and maintained as a list provided the velocity at the new locations is higher than the original ones. $p6$ is the farthest point on the robot's trajectory visible from its current location at A

4. On this set of locations $A, p1, p2, p3, p4, p5, p6$ starting from the current location at A, find a trajectory sequence shorter in time than the current sequence of $A, B, p6$ if it exists.
5. The steps 1 to 4 are repeated until the robot reaches the target.

It should be noted that if a collision with an object is detected, a collision free location is first found that connects the current location with another location on the original trajectory and this new collision free path is further adapted for a time-reduced path if it exists. Also, while the velocities are computed over a distance d_{safe}, that part of the remaining trajectory that is visible from the current location is considered for adapting to a better time-length.

4.5 Analysis and Results at the Planning Stage

In this section the results of incorporating the velocity profile computation as a consequence of considering robot and environment dynamics and sensor capacities at the planning stage and the subsequent adaptation of paths to better time of trajectory is analyzed. Figure 4.12-a shows the path computed by a typical holonomic planner [9] and its corresponding velocity profile. The velocity corresponding to the robot's location on the trajectory (shown as a small circle) is marked by a straight line labeled m on the profile. The dark star-shaped polygon centered at the robot depicts the visibility of the robot at that instant and is called the visibility polygon. The figure is a snapshot of the instant when the robot begins to decelerate to a velocity less than half the current velocity as it closes down on the vertex a marked in the figure. Evidently from the visibility polygon the vertex a casts a shadow and the closer the robot gets to it, the slower the velocity must be.

Figure 4.12-b is the time reduced counterpart of Figure 4.12-a. The snapshot is once again at a location close to vertex a. Staying away from a permits nearly

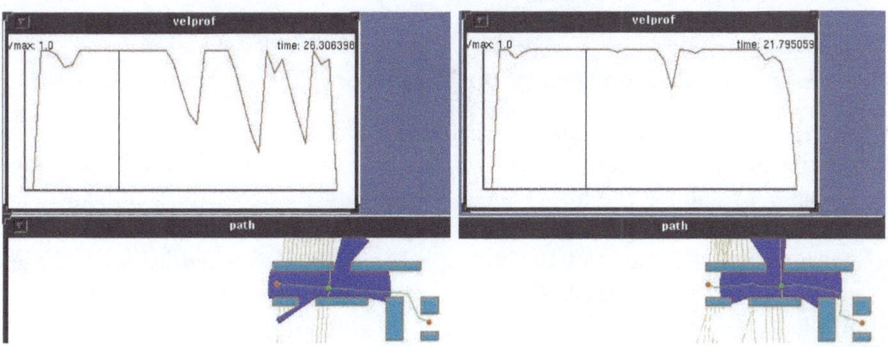

Fig. 4.12. (a) Path computed by a typical planner and its velocity profile shown on the top. The robot's velocity corresponding to its location on the trajectory is shown by a vertical line on the profile and labeled as m and (b) Path obtained after adaptation to reduced time-length.

maximum velocity. The dip observed in the profile due to vertex a is negligible. Similarly staying away from other vertices such as b allows for a trajectory time of $21.79s$ compared to $26.30s$ for Figure 4.12-a. Modification of the trajectory for shorter time proceeds along the lines of Section 4.3.4. For the two examples discussed, the robot's maximum acceleration and deceleration was fixed at $1m/s^2$, maximum velocity at $1m/s$ and the sensor range at $7m$. The maximum bound on the object's velocities was $1.5m/s$. Figure 4.13 depicts the planned trajectory and velocity profiles before (left) and after (right) reduction of trajectory time for our laboratory environment. The time reduced trajectory is shorter by more than 8 seconds as it widens its field of view by moving away from the bends while turning around them.

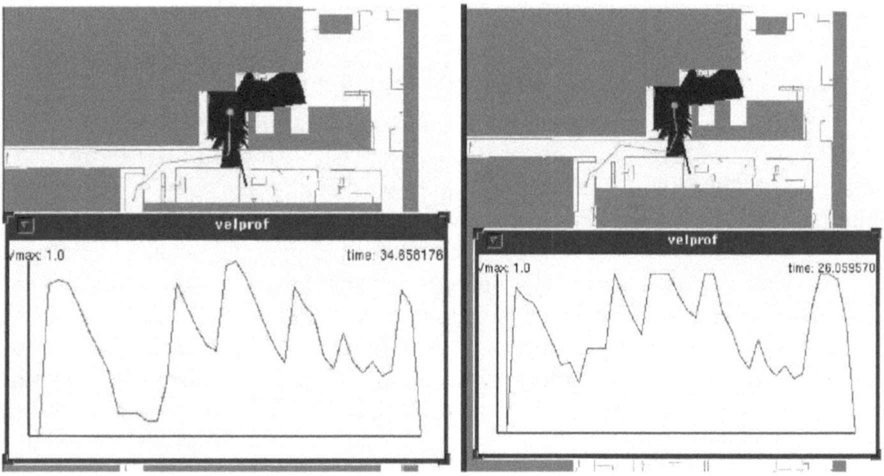

Fig. 4.13. (a) Planned trajectory before adaptation to a reduced time and (b) Time reduced trajectory at planning stage

4.5.1 Effect of Remembering on Trajectory Time

Figure 4.14 shows an environment with four corridors with a path obtained by minimizing time. It also portrays the robot's field of vision as it enters a corridor. The velocity profile for the above path is shown in Figure 4.15. The location of the robot corresponding to its location in Figure 4.14 is shown through the vertical line. The locations of the robot as it decelerates when its field of view of each of the corridors vanishes is also marked with the respective numbers on the profile.

Though the path of Figure 4.14 is minimized in time its velocity profile still shows decelerations in the vicinity of the corridors. This is due to the phenomenon discussed in Section 4.3.5 where the robot becomes blind to many parts of the environment it had seen at the preceding instant. However, when the robot is able to remember the previous images, the need to decelerate is

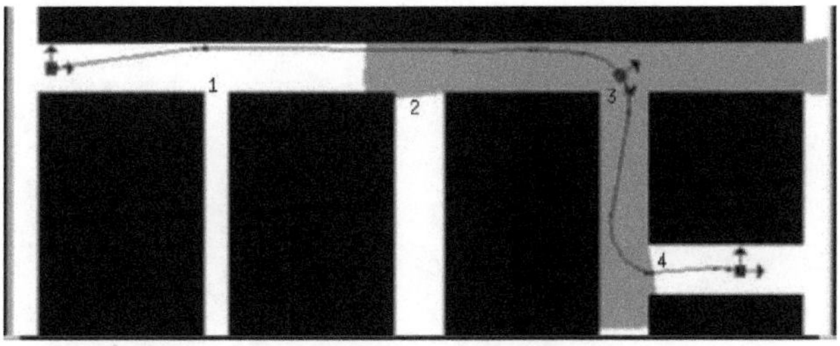

Fig. 4.14. Computed path in an environment with four corridors

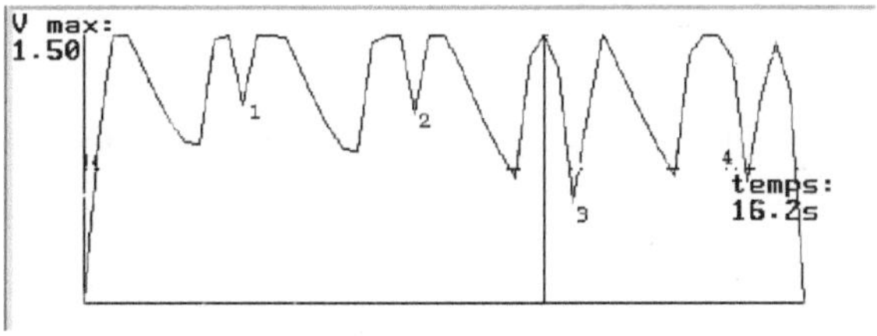

Fig. 4.15. Velocity profile for Figure 4.14. Position of the robot shown in vertical line. Decelerations near the corridors are also marked with the same numbers

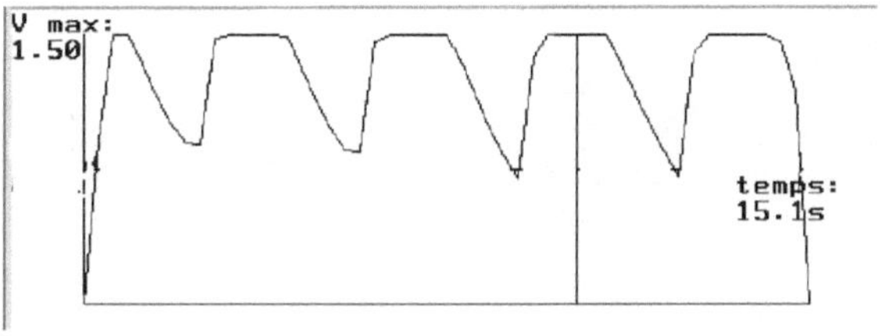

Fig. 4.16. The velocity profile obtained after incorporation of memory

nullified and the trajectory time is further reduced. Figure 4.16 illustrates this where the decelerations shown in the velocity profile of Figure 4.15 at locations 1, 2, 3 and 4 are now absent.

4.6 Simulation and Experimental Results

4.6.1 Velocity Profiles During Planning and Execution

In this section the velocity profiles obtained during planning and execution stages are compared in the absence of any new objects during execution. Figure 4.17-a shows a simple planned trajectory and the corresponding velocity profile for our lab environment. Some of the obstacles are filled in gray and others are shown as segments (in gray). The robot is shown as a small circle and the star shaped

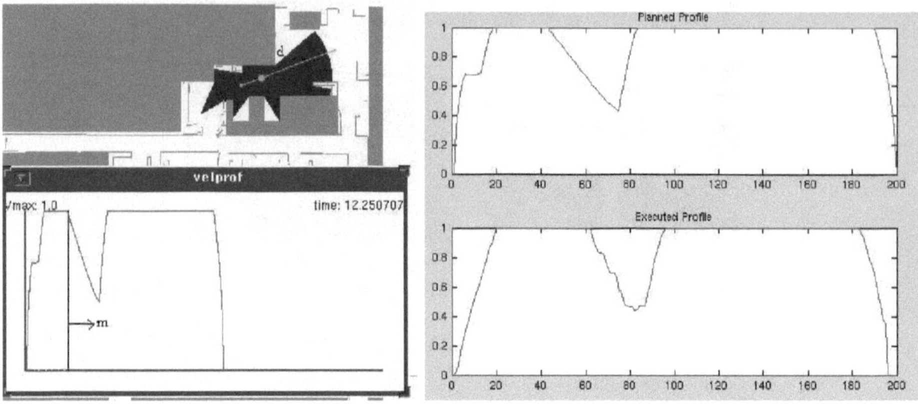

Fig. 4.17. (a) A simple planned trajectory and its velocity profile and (b) The planned and executed velocity profile in simulation

Fig. 4.18. The Nomad XR4000 used in our experiments

polygon in black represents the field of vision of the robot at that location. The vertical line, marked m in the velocity profile represents the velocity of the robot corresponding to its position on the trajectory. The profile shows a subsequent drop in velocity, a consequent of robot getting closer to region marked, d, to which it is blind. Figure 4.17-b compares the planned and executed (in simulation) velocity profile. The executed trajectory tallied to a time of $12.28s$ in comparison with $12.25s$ for the planned profile. These figures illustrate that the executed profiles and execution times are close to the planned profiles and times while there are no changes in the environment.

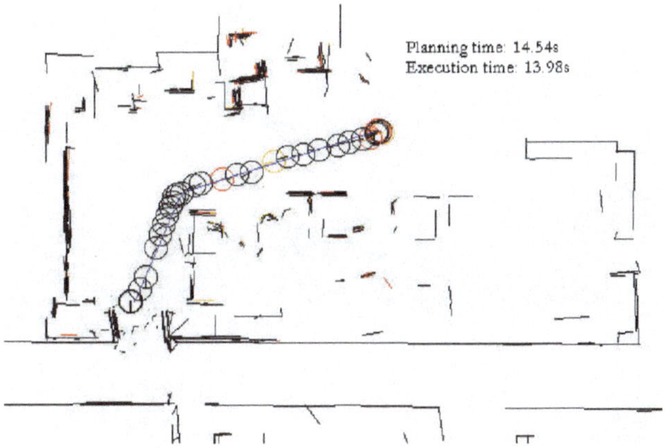

Fig. 4.19. Execution of the original planned path by the Nomad

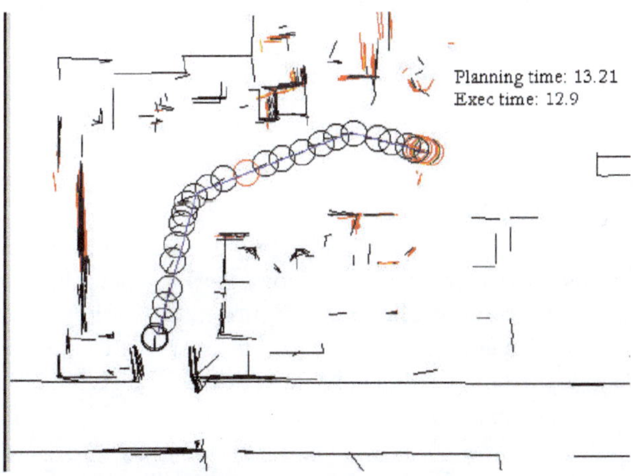

Fig. 4.20. Execution of the time-reduced path by the Nomad

Figures 4.19 and 4.20 show the execution by the Nomad $XR4000$ (Fig. 4.18) of paths computed by a standard planner. Figure 4.19 corresponds to the original path computed by the planner and Figure 4.20 is its time reduced counterpart. The velocity profiles during execution of the two paths are shown in Figure 4.21. Some of the bigger drops in the un-reduced profile are absent in the reduced profile as the robot avoids turning close to the obstacles that form the bends. The path of Figure 4.20 got executed in $12.9s$ while the path in Figure 4.19 was executed in $13.98s$. The figures are meant as illustrations of the theme that trajectories deformed to shorter time-lengths at planning stage are also executed in shorter time during implementation than their un-reduced versions.

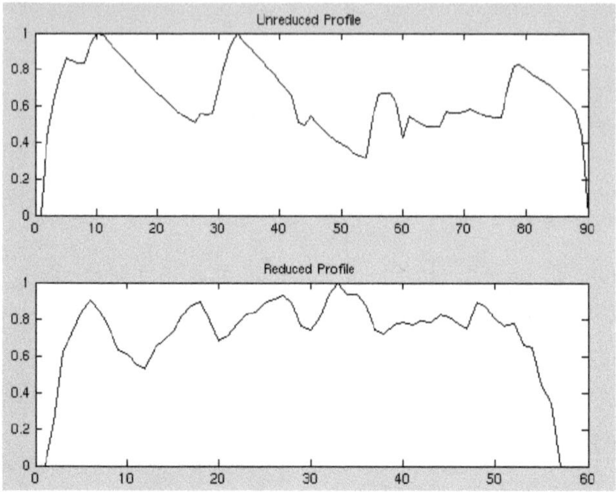

Fig. 4.21. The top profile corresponds to the path executed in Figure 4.19 and the bottom to Figure 4.20

4.6.2 Online Adaptation of Paths for Better Trajectory Time

This section presents results of the algorithm in the presence of newly added objects that affect the velocities of the robot in real-time. Figure 4.22-a shows a path where the robot avoids the two new segments $S1$ and $S2$ intersecting the original planned trajectory but does not adapt its path for better time. The velocity profile for the same is shown in Figure 4.22-b. Figure 4.23 is the counterpart of Figure 4.22 where the robot adapts its path to a better time-length reactively. The big dips in the velocity profile of Figure 4.22-b are filtered in

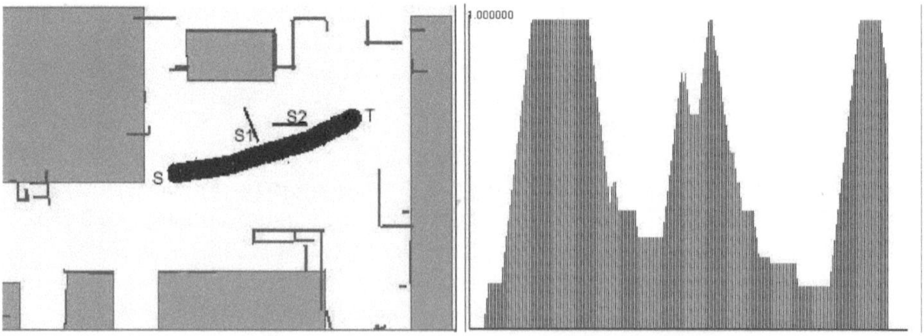

Fig. 4.22. (a) A simulated execution in the presence of two new segments $S1$ and $S2$ along with the corresponding velocity profile. The path is not adapted to better time-length. Start and goal locations marked as S and T and (b) The velocity profile for the execution of Figure 4.22.

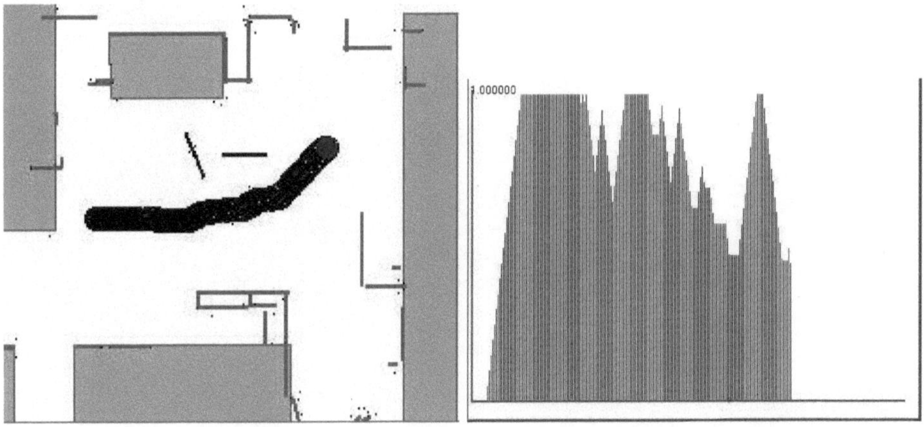

Fig. 4.23. (a) Path of Figure 4.22 adapted to better time-length and (b) The corresponding velocity profile

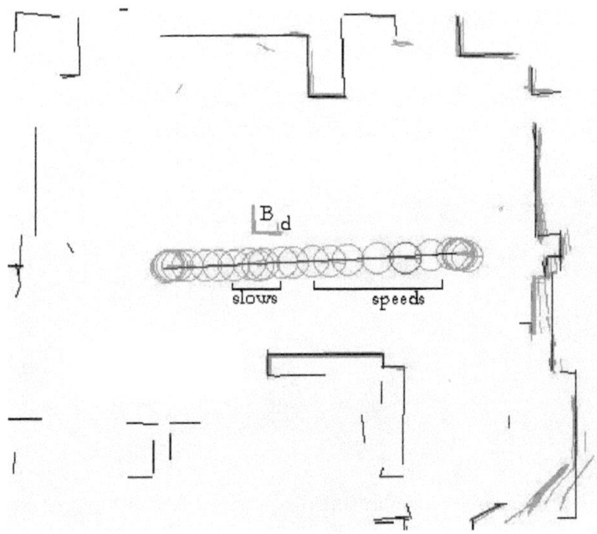

Fig. 4.24. Un-reduced path executed by the Nomad $XR4000$. The vertex d of the new box shaped object B forces a slow down near it.

Figure 4.23-b considerably as the robot avoids the obstacles with larger separation. The time reduced execution tallied to $10.9s$ while the un-reduced version was executed in $12.5s$. The trajectory time at planning was $7.9s$. The above graphs are those obtained in simulation.

Figure 4.24 shows the un-reduced executed path by the $XR4000$ Nomadic robot in our laboratory at LAAS. The obstacles in the original map are shown by black lines, while the segments perceived by the SICK laser are shown in

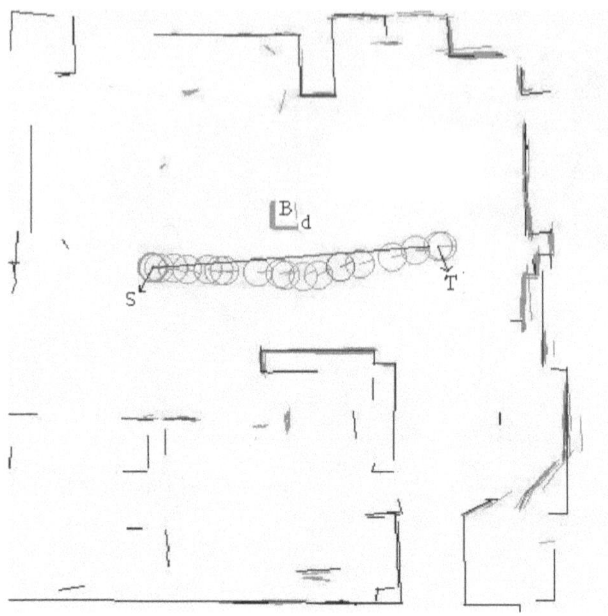

Fig. 4.25. Time reduced path executed by the Nomad $XR4000$. Increasing linear and angular separation from vertex d facilitates a higher speed.

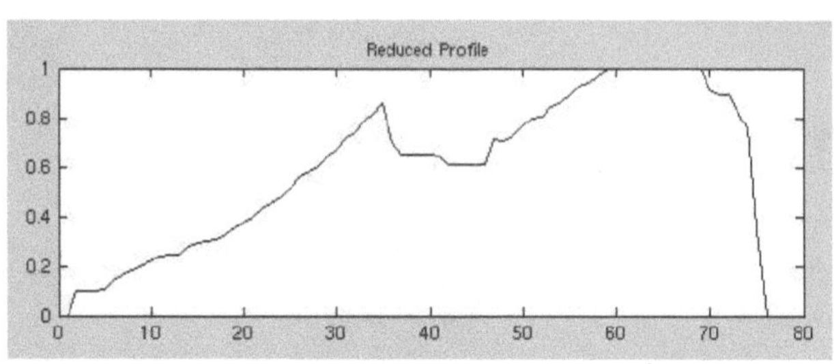

Fig. 4.26. Velocity profile for the path executed by the Nomad in Figure 4.25

lighter shades of gray. Some of these segments get mapped to the ones in the map and the others are considered new segments. This is done by a segment based localization algorithm. The segments of concern here are those which form a box shaped obstacle marked B in Figure 4.24. The vertex d of this obstacle casts a shadow on the robot's sensory field, which forces it to slow down at those locations due to Equation 4.2. The execution time for this un-reduced path is $10.6s$. The time reduced counterpart is shown in Figure 4.25 that tallied to $9.6s$. The original planning time was $8.8s$ in the absence of the box shaped object. The velocity profile for the same is shown in Figure 4.26.

4.7 Conclusions and Scope

A proactive safe planning algorithm and its reactive version that facilitates real-time execution has been presented. The proactive nature of the algorithm stems from the computed velocity profile, $v_\tau(s)$, that guarantees immobility of the robot before collision with any of the possible mobiles that could interfere its future trajectory from regions blind to its sensor. The proactivity does not however come at the cost of robot's velocity or trajectory time. The knowledge of $v_\tau(s)$ computed over the trajectory $\tau(s)$ further facilitates reduction of the over all trajectory time $T(\tau)$ by adaptation of the initially planned path. Analysis of the scheme at the planning stage depict that the robot can have a velocity profile that achieves its maximum possible velocity for a sustained duration without many dips provided it stays away from doorways and narrow passages along its path. Memory of previously observed scenes also enhance the robot's performance through reduced trajectory time and a more uniform velocity profile.

A reactive extension of the scheme that facilitates real-time simulation and implementation is also presented. The scheme maintains the underlying philosophy of computing safe velocities and modification of paths for better trajectory time. Simulation and experimental results at real-time corroborate our earlier results obtained at the planning stage (that by keeping away from vertices of objects that could hide mobiles the robot could move at higher velocities and obtain better time-lengths) and thus the efficacy of overall strategy is vindicated. The minimum distance over which the velocities need to be computed on the remaining trajectory during real-time such that the computed velocities are safe is theoretically established. This avoids repetitive computation of velocities over the entire remaining trajectory for every motion command, thereby reducing computational intensity and facilitating for real-time implementation. The methodology could be useful in the context of personal robots moving in areas where interference with mobile humans especially aged ones are generally expected.

Immediate scope of this work involves incorporating memory of phenomena at the reactive level such that higher speeds are possible. The methodology needs to be validated in the presence of mobile objects that actually impinge on the path from blind zones with a provision for the robot to avoid the objects without halting continuing to respect safety considerations as well as minimizing trajectory time.

Acknowledgments

The work described in this paper was conducted within the EU Integrated Project COGNIRON ("The Cognitive Companion") and was funded by the European Commission Division FP6-IST Future and Emerging Technologies under Contract FP6-002020 and by the French National Program ROBEA.

References

1. R. Alami, T. Simeon, and K.Madhava Krishna. On the influence of sensor capacities and environment dynamics onto collision-free motion plans. *IEEE/RSJ International Conference on Intelligent Robots and Systems, EPFL, Swizerland*, 2002.
2. J.C. Alvarez, A. Skhel, and V. Lumelsky. Accounting for mobile robot dynamics in sensor-based motion planning: experimental results. *IEEE International Conference on Robotics and Automation, Leuven (Belgium)*, 1998.
3. B. Bouily, T. Simeon, and R. Alami. A numerical technique for planning motion strategies of a mobile robot in presence of uncertainty. *IEEE International Conference on Robotics and Automation, Nagoya (Japan)*, 1995.
4. D. Cruzel. Planification de mouvements sous contraintes de perception. Master's thesis, LAAS-CNRS, 1998.
5. P. Fiorinin and Z. Schiller. Motion planning in dynamic environments using velocity obstacles. *International Journal of Robotics Research*, 17(7):760–772, 1998.
6. S. Fleury, P. Soueres, and J.P. Laumond. Primitives for smoothing mobile robot trajectories. *IEEE Transactions on Robotics and Automation*, 11(3):441–448, 1995.
7. M. Khatib, B. Bouily, T. Simeon, and R. Chatila. Indoor navigation with uncertainty using sensor-based motions. *IEEE International Conference on Robotics and Automation, Albuquerque (USA)*, 1997.
8. K.Madhava Krishna, R. Alami, and T. Simeon. Moving safely but not slowly - reactively adapting paths for better trajectory times. *IEEE International Conference on Advanced Robotics, Quimbra, Portugal*, 2003.
9. J.C. Latombe. *Robot Motion Planning.* Kluwer Academic, 1991.
10. A. Lazanas and J.C. Latombe. Motion planning with uncertainty: a landmark approach. *Artificial Intelligence*, pages 287–315, 1995.
11. J. Minguez and L. Montano. Nearness diagram navigation. a new real-time collision avoidance approach. *IEEE/RSJ International Conference on Intelligent Robots and Systems*, 2000.
12. N. Roy and S. Thrun. Motion planning through policy search. *IEEE/RSJ International Conference on Intelligent Robots and Systems, EPFL, Swizerland*, pages 2419–2425, 2002.
13. Z. Schiller, F. Large, and S. Sekhavat. Motion planning in dynamic environments: Obstacles moving along arbitrary trajectories. *IEEE International Conf. on Rob. Automat.*, pages 3716–3721, 2001.
14. C. Strachniss and W. Burgard. An integrated approach to goal-directed obstacle avoidance under dynamic constraints for dynamic environments. *IEEE/RSJ International Conference on Intelligent Robots and Systems, EPFL, Swizerland*, 2002.
15. S. Suri and J. O'Rourke. Worst-case optimal algorithms for constructing visibility polygons with holes. *ACM Symp. on Computational Geometry*, 1986.

5

Motion Planning in Dynamic Environments

Zvi Shiller[1], Frederic Large[2], Sepanta Sekhavat[2], and Christian Laugier[2]

[1] Department of Mechanical Engineering-Mechatronics
College of Judea and Samaria
Ariel, Israel
shiller@yosh.ac.il

[2] INRIA Rhone-Alpes
38330 Grenoble, France
laugier@imag.fr

Summary. Motion planning in dynamic environments is made possible using the concept of *velocity obstacles.* It maps the colliding velocities of the robot with any moving or static obstacle to the robot's velocity space. Collision avoidance is achieved by selecting the robot velocity outside the velocity obstacles. This concept was first proposed in [3] for the linear case of obstacles moving on straight line trajectories, and is extended here to obstacles moving along arbitrary trajectories. The *non-linear velocity obstacle* (NLVO) takes into account the shape, velocity and path curvature of the moving obstacle. It allows to select a *single* avoidance maneuver (if one exists) that avoids any number of obstacles that move on any known trajectories. The nonlinear v-obstacle can be generated as a time integral of the colliding velocities, or by computing its boundaries using analytic expressions.

5.1 Introduction

Dynamic environments represent an important and a growing segment of modern automation, in applications as diverse as, intelligent vehicles negotiating freeway traffic, air and sea traffic control, automated wheel chairs [2], automated assembly, and animation. Common to these applications is the need to quickly select avoidance maneuvers that avoid potential collisions with moving obstacles.

There has been a sizable body of work on this problem (see [3] for an extended survey). However, most have relied explicitly on the evolutions of the positions of both robot and obstacles to determine future potential collisions. A more efficient approach that determines future potential collision based on the current velocities of both robot and obstacles was presented in [3]. It is based on the concept of Velocity Obstacles (v-obstacles), which represent the sets of the robot's colliding velocities with each moving (or static) obstacle. The v-obstacles allow to efficiently select a *single* velocity by the robot that avoids any number of moving obstacles (if such a velocity exists), given that they maintain their current velocities. This method is termed *first order* since it is based on the first order approximation (velocity) of the trajectories followed by the moving obstacles. A similar first order approach for collision detection between moving

C. Laugier and R. Chatila (Eds.): Autonomous Navi. in Dyn. Environ., STAR 35, pp. 107–119, 2007.
springerlink.com

obstacles of arbitrary shapes was developed separately in [1] based on results from missile guidance.

Although only first order, it was shown to successfully handle obstacles moving on arbitrary trajectories [3]. However, if the obstacles are moving along general trajectories, then deriving the v-obstacle to reflect their exact trajectories can result in fewer adjustments by the avoiding vehicle. In addition, the exact (non-linear) v-obstacle may be necessary in cases where the linear v-obstacle indicates collision, when in fact there is none because of the curved motion of the other vehicle. For example, a vehicle moving along a curved road, turning right, appears to be on a collision course with the vehicles on the opposite lane, using the linear v-obstacle, when in fact there is no imminent collision if the approaching vehicles stay on the curved lane. The v-obstacle that takes into account the road curvature should eliminate this confusion. Another example is the problem of merging with traffic along a turn-around, where the vehicles moving around the curve appear to be on a collision course with the vehicle waiting to join, using the linear v-obstacle. Here too, the appropriate v-obstacle will allow the waiting vehicle to plan an efficient maneuver that safely merges with the traffic flow.

This paper treats obstacles that move along arbitrary trajectories. It focuses on the unified representation of static and moving obstacles by v-obstacles, and not on the avoidance strategy problem. This representation assumes a single avoiding intelligent agent and any number of apathetic (neither competitive nor cooperative) obstacles.

The paper is organized as follows: we first review the *linear* velocity obstacle in Section 2. Section 3 extends the velocity obstacle to general trajectories, and Section 4 presents a simple approximation of its boundaries. Section 5 presents several examples.

5.2 Linear V-Obstacle

The v-obstacle represents the colliding velocities of a robot, or equivalently, the "velocity obstacle," at a given time, which are mapped to the configuration space. The geometry of this set can be precisely and easily defined, as discussed below. For simplicity, we consider circular robots and obstacles. Growing the obstacles by the radius of the robot transforms the problem to a point robot that avoids circular obstacles. It is also assumed that the instantaneous states (position, velocity, and acceleration) of obstacles moving along arbitrary trajectories are either known or measurable.

A few words about notation: henceforth, A denotes a point robot, and $\mathcal{B}$ denotes the set of points defining the geometry of and obstacle. Since the obstacle is solid, $\mathcal{B}$ does not depend on t. $B(t)$ denotes the set of points occupied by the obstacle B at time t. Thus, if $b \in \mathcal{B}$ is some representative point (usually the center) of $\mathcal{B}$, and it coincides with the trajectory $c(t)$ at time t, then

$$B(t) = c(t) + \mathcal{B}. \tag{5.1}$$

a/b denotes a *ray* that consists of the half line that originates at a, passes by b, and does not include a. Similarly, a/v denotes a ray that originates at a and is parallel to v.

The linear v-obstacle [3] is demonstrated for the scenario shown in Figure 5.1, where, at time t_0, obstacle $B(t_0)$, moving at some linear velocity v_b, is to be avoided by a point robot A. The linear v-obstacle at time t_0 is constructed by first generating the so called *relative velocity cone* (RVC) by sweeping a half line from A along ∂B, the boundary of $B(t_0)$. RVC is thus defined as the union of all rays originating from A and passing through $\partial B(t_0)$:

$$RVC = \cup A/b, \quad b \in \partial B(t_0). \tag{5.2}$$

RVC is the set of all velocities, $v_{a/b} \neq 0$, of A *relative* to B that would result in collision at some time $t \in (0, \infty)$, assuming that the obstacle stays on its current course at its current speed. Consequently, relative velocities outside of RVC ensure avoidance of B at all times under the same assumptions. Translating RVC by v_b produces the *velocity obstacle*, VO [3]:

$$VO = v_b + RVC \tag{5.3}$$

where in this context, "+" denotes the Minkowski sum:

$$VO = \{x | x = y + v_b, y \in RVC\}. \tag{5.4}$$

Thus, VO represents a set of *absolute* velocities of A, v_a, that would result in collision at some time $t \in (0, \infty)$. Geometrically, each point $x \in VO$ represents a vector originating at A and terminating at x. In Figure 5.1, v_{a2} is a colliding velocity, whereas v_{a1} is not. Note that VO reflects the positions of A and B, and v_b, at $t = t_0$, but is independent of v_a.

We call VO a *linear* v-obstacle because it ensures avoidance using a *single* maneuver at time t_0 under the assumption that the obstacle maintains its current course and speed. For an obstacle moving on a curved path, its current velocity represents a first order approximation of its actual trajectory at time t_0. Similarly, the linear v-obstacle represents a first order approximation of the non-linear v-obstacle presented later.

By construction, RVC is a convex cone with extreme rays tangent to $B(t_0)$, as was shown in Figure 5.1. The extreme rays represent the set of relative velocities that would result in A grazing $B(t)$ at some future time. The points of contact between A and $B(t)$ are the tangency points b_l and b_r between the extreme rays of RVC and $B(t_0)$, as shown in Figure 5.1 [6]. The time of contact is determined by the magnitude of $v_{a/b}$.

5.2.1 A Temporal Linear V-Obstacle

VO represents the velocities of A that would result in collision at any time $t = (0, \infty)$. It is useful to identify a subset of VO that would result in collision at a specified time.

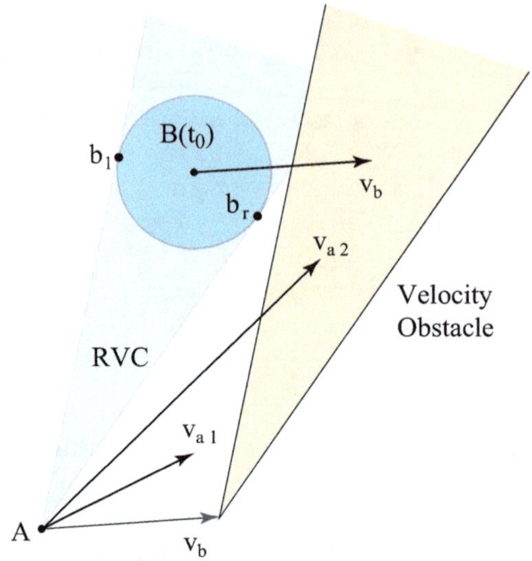

Fig. 5.1. The Linear Velocity Obstacle

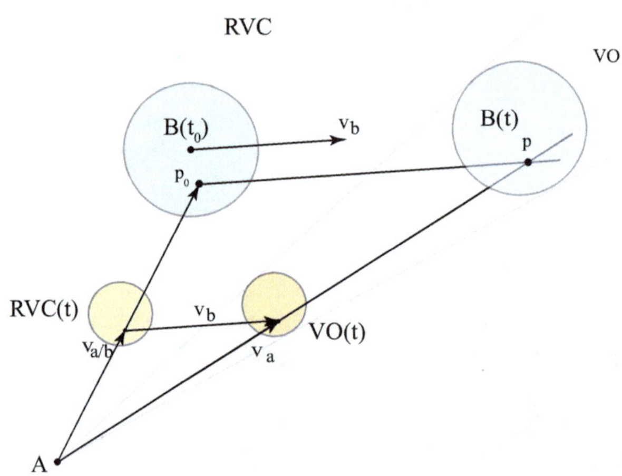

Fig. 5.2. A Temporal Element of VO

Traveling at some relative velocity $v_{a/b} \in RVC$ over $t = [t_0, t]$, A will reach the point $p \in B(t)$ along the ray defined by $A(t_0)$ and v_a, as shown in Figure 5.2:

$$p = (t - t_0)v_a \tag{5.5}$$

where $p \in \mathcal{R}^2$ is defined in a coordinate frame centered at $A(t_0)$. Solving for the absolute velocity v_a that would reach p at time t yields:

$$v_a = \frac{1}{(t - t_0)}p. \tag{5.6}$$

Expressing p as

$$p = p_0 + v_b(t - t_0) \tag{5.7}$$

where $p_0 \in B(t_0)$, yields

$$v_a = \frac{1}{(t-t_0)}(p_0 + v_b(t-t_0)) = v_b + \frac{1}{(t-t_0)}p_0. \tag{5.8}$$

Substituting p_0 with the set of all points in $B(t_0)$ yields the set of all absolute velocities that would result in collision with any point of $B(t)$ at time $t > t_0$:

$$VO(t) = v_b + RVC(t), \tag{5.9}$$

where

$$RVC(t) = \frac{1}{(t-t_0)}B(t_0) = \frac{c(t_0) + \mathcal{B}}{t - t_0}, \tag{5.10}$$

where $c(t_0)$ is the location of the obstacle's center at time t_0. The shape of $RVC(t)$ depends on $\mathcal{B}$ and its relative position to A at time t_0, as shown in Figure 5.2.

The linear v-obstacle VO is thus defined as the union of the temporal elements:

$$VO = \bigcup_{t>t_0} VO(t), \tag{5.11}$$

or

$$VO = v_b + \bigcup_{t>t_0} \frac{c(t_0) + \mathcal{B}}{t - t_0}. \tag{5.12}$$

Clearly, the linear v-obstacle can be truncated to reflect collisions within a specified time interval $[t_1, t_2]$ simply by defining the time interval in (5.12):

$$VO(t_1, t_2) = v_b + \bigcup_{t_2 \geq t > t_1} \frac{c(t_0) + \mathcal{B}}{t - t_0}. \tag{5.13}$$

Truncating VO to reflect specific time limits allows to focus the motion planning problem on potential collisions occurring within a specified time interval, such as imminent collisions occurring within some given time horizon [3]. This permits avoidance maneuvers that are potentially risky, but at times that are practically insignificant at the decision time t_0. It also allows the consideration of large static obstacles, such as surrounding walls and highway barriers, whose v-obstacles cover the entire velocity space.

5.3 The Non-linear V-Obstacle

The non-linear v-obstacle ($NLVO$) applies to the scenario shown in Figure 5.3, where, at time t_0, a point robot, A, attempts to avoid a circular obstacle, $\mathcal{B}$, that at time t_0 is located at $c(t_0)$, and is following a general known trajectory, $c(t)$. $NLVO$ thus consists of all velocities of A at t_0 that would result in collision with the obstacle at any time $t > t_0$. Selecting a *single* velocity, v_a, outside $NLVO$ should therefore avoid collision at all times, or

$$(A + v_a^{t_0} t) \notin (c(t) + \mathcal{B}) \quad \forall t > t_0 \tag{5.14}$$

where $v_a^{t_0}$ denotes the velocity of A at time t_0. Note that the emphasis here is on selecting a *single* avoidance maneuver at time t_0, for any trajectory followed by the obstacle.

The non-linear v-obstacle is constructed by first determining the absolute velocities of A, v_a, that would result in collision at a specific time t. Referring to Figure 5.3, $v_a^{t_0}(t,p)$ that would result in collision with point $p \in B(t)$ at time $t > t_0$ expressed in a frame centered at $A(t_0)$ is simply

$$v_a^{t_0}(t,p) = \frac{c(t) + r_p}{t - t_0}, \tag{5.15}$$

where r_p is the vector to point p in the obstacle's fixed frame.

Similarly, the set, $NLVO(t)$ of all absolute velocities of A that would result in collision with any point in $B(t)$ at time $t > t_0$ is:

$$NLVO(t) = \frac{c(t) + \mathcal{B}}{t - t_0} \tag{5.16}$$

The nonlinear temporal v-obstacle (5.16) is similar to the linear case (5.10), except that here $c(t)$ changes with time, as opposed to $c(t_0)$ used in (5.10).

Geometrically, $NLVO(t)$ is a scaled $\mathcal{B}$, bounded by the cone formed between A and $B(t)$. Note that the extreme rays of this cone are not necessarily the points where A grazes B, as discussed later. Note also that $NLVO(t)$ is independent of $v_b(t)$, since it is a function of $B(t)$ and not of its future positions (determined by $v_b(t)$). Similarly to the linear case, the non-linear v-obstacle, $NLVO$, of obstacle B that is moving along a general trajectory $c(t), t = (t_0, \infty)$ is the union of its temporal elements:

$$NLVO = \bigcup_{t>t_0} \frac{c(t) + \mathcal{B}}{t - t_0}, \tag{5.17}$$

The non-linear v-obstacle represents the set of all absolute velocities of A at time $t = t_0$ that would collide with $B(t)$ at time $t = (t_0, \infty)$. It is a warped cone with apex at A. Its center line consist of the obstacle's trajectory $c(t)$ divided by $t - t_0$, which as $t \to \infty$ approaches A.

Similarly to the linear case, the v-obstacle can be truncated by some time horizon to focus the motion planning problem on potential collisions occurring within a specified time interval. Not considering the time horizon would dictate the consideration of collisions occurring at any time, which would be too prohibitive and would preclude safe maneuvers that are only momentarily dangerous. The minimum time horizon is the time it would take for the robot to clear the way of the obstacle. For a static obstacle, this is simply the time of an optimal avoidance maneuver, which for a typical passenger car is approximately 1 s [7]. This implies that assuming a time horizon shorter than 1 s would leave

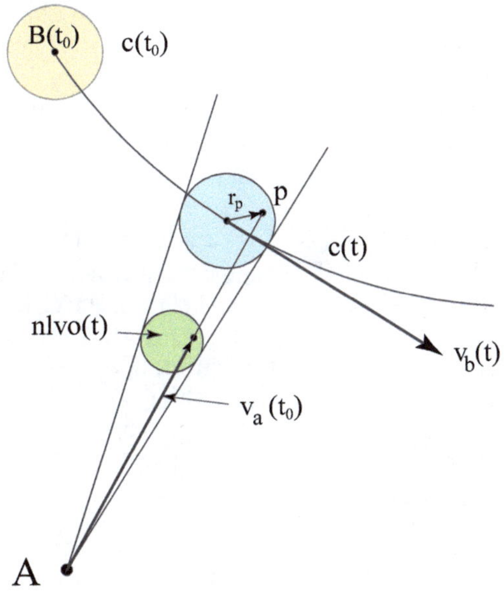

Fig. 5.3. Construction of the non-linear v-obstacle

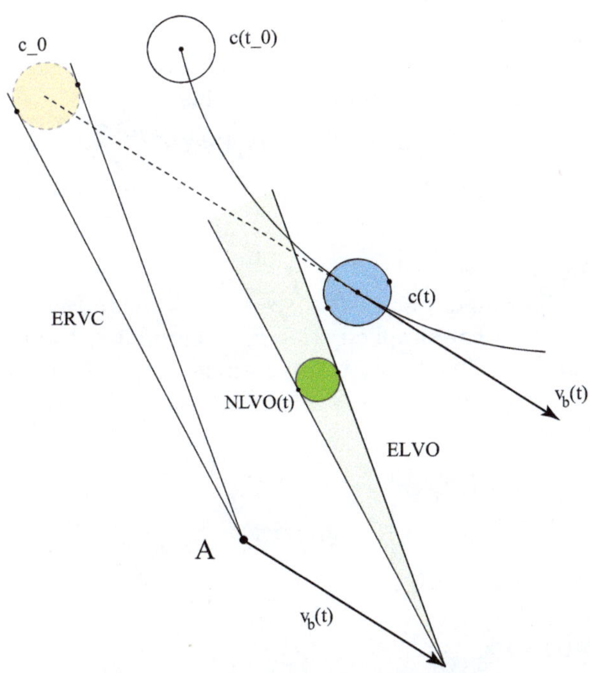

Fig. 5.4. The Equivalent Linear V-Obstacle

no time for the approaching vehicle to avoid the static obstacle. Using a larger time horizon would be safe, but would preclude safe avoidance maneuvers.

The boundaries of $NLVO$ represent velocities that would result in A grazing $\mathcal{B}$. The tangency points between A and $B(t)$, assuming that $B(t)$ does not rotate relative to the path coordinates, are determined from the *equivalent linear v-obstacle*, $ELVO(B,t)$ [6], which is the linear v-obstacle of a *virtual* $\mathcal{B}$ that reaches $c(t)$ along a straight line at a constant $v_b(t)$ over $[t_0, t]$.

Referring to Figure 5.4, $ELVO(B,t)$ is constructed by integrating $v_b(t)$ backwards in time to t_0, constructing $ERVC$ (the equivalent relative velocity cone) at point $c_0 = c(t) - v_b(t)(t - t_0)$, then translating $ERVC$ by $v_b(t)$. Similarly to the linear v-obstacle, the points where A grazes $B(t)$ are the tangency points between $ERVC$ and the virtual $\mathcal{B}$ at t_0 [6]. These points correspond to the tangency points between $ELVO$ and $NLVO(t)$, as shown in Figure 5.4. The tangency points on $NLVO(t)$ correspond to the velocities of A that would graze $B(t)$. The $NLVO$ can thus be computed using (5.17), or by approximating the tangency points, as discussed next.

5.4 Approximating the Boundary of $NLVO$

A conservative representation of the boundary of $NLVO$ can be derived using complex numbers. The trajectory, $c(t)$, followed by the obstacle, $\mathcal{B}$, can be represented by

$$c(t) = d(t)e^{i\theta(t)}. \tag{5.18}$$

where $i = \sqrt{-1}$, and $d(t)$ and $\theta(t)$ are measured in a coordinate frame centered at A (see Figure 5.5). Differentiating (5.18) yields $v_b(t)$:

$$v_b(t) = \dot{d}(t)e^{i\theta(t)} + i\dot{\theta}(t)d(t)e^{i\theta(t)}. \tag{5.19}$$

Assuming $t_0 = 0$, we divide (5.18) by t to obtain the center, $c_v(t)$, of the $NLVO$:

$$c_v(t) = \frac{d(t)}{t}e^{i\theta(t)}. \tag{5.20}$$

As stated earlier, the boundary of $NLVO$ consists of the tangency points between $ELVO$ and $NLVO(t)$. For simplicity, let $\mathcal{B}$ be circular, although the following applies, with minor modifications, to $\mathcal{B}$ of any shape. $ELVO$ is a cone with a center line passing through $c_v(t)$ and an apex at $v_b(t)$. The center line, $c_l(t)$, of this cone is therefore

$$\begin{aligned} c_l(t) &= c_v(t) - v_b(t) \\ &= [(\frac{1}{t} - i\dot{\theta}(t))z(t) - \dot{z}(t)]e^{i\theta(t)}. \end{aligned} \tag{5.21}$$

The corresponding unit vector is

$$\begin{aligned} \hat{c}_l(t) &= \frac{c_l(t)}{\|c_l(t)\|} \\ &= \frac{[(1 - it\dot{\theta}(t))d(t) - t\dot{d}(t)]e^{i\theta(t)}}{\sqrt{(1+\dot{\theta}^2(t))d^2(t)t^2 + \dot{d}^2(t)t^2 - 2d(t)\dot{d}(t)t}}. \end{aligned} \tag{5.22}$$

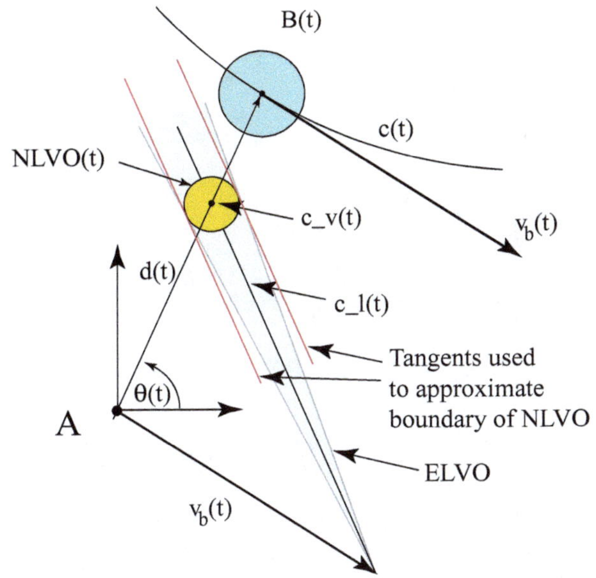

Fig. 5.5. Approximating the boundaries of $NLVO$

The exact boundary of $NLVO$ is traced by the tangency points between $ELVO$ and $NLVO(t)$. For simplicity, we approximate the exact tangency points by a simpler procedure that computes instead the tangency points, $vo_r(t)$ and $vo_l(t)$ between $NLVO(t)$ and the two tangents passing parallel to $\hat{c}_l(t)$, as shown in Figure 5.5:

$$vo_r(t) = c_v(t) + i\frac{r}{t}\hat{c}_l(t) \tag{5.23}$$

$$vo_l(t) = c_v(t) - i\frac{r}{t}\hat{c}_l(t), \tag{5.24}$$

where r is the radius of $\mathcal{B}$. The points $vo_r(t)$ and $vo_l(t)$ are exterior to the true $NLVO$ because $(b_r, b_l) \cap (vo_r(t), vo_l(t))$, where (a, b) represent the boundary arc between a and b. Thus, using $vo_r(t)$ and $vo_l(t)$ yields a conservative representation of $NLVO$, except near singularity points, where $A \in NLVO(t)$ [6].

Assuming a circular obstacle of radius r that moves along a circular trajectory of radius R, centered at point $O = De^{i\phi}$ in a coordinate frame centered at A, at angular speed ω, starting at $\theta(t_0) = \theta_0$, we obtain the following expressions for the center and boundaries of the corresponding $NLVO$:

$$c(t) = De^{i\phi} + Re^{i\theta(t)} \tag{5.25}$$

$$c_v(t) = \frac{1}{t}[De^{i\phi} + Re^{i\theta(t)}] \tag{5.26}$$

$$v_b(t) = i\omega Re^{i\theta(t)} \tag{5.27}$$

$$c_l(t) = \frac{1}{t}[De^{i\phi} + (1 - i\omega t)Re^{i\theta(t)}] \tag{5.28}$$

$$\hat{c}_l(t) = \frac{De^{i\phi} + (1 - i\omega t)Re^{i\theta(t)}}{\sqrt{D^2 + R^2 + (\omega tR)^2 + 2DR(cos\alpha + \omega t sin\beta)}} \tag{5.29}$$

$$\theta(t) = \omega t + \theta_0 \tag{5.30}$$

$$\alpha = \theta(t) - \phi \tag{5.31}$$

$$\beta = \theta(t) - \phi \tag{5.32}$$

$$vo_r(t) = \frac{1}{t}[De^{i\phi} + (1 - i\omega t)Re^{i\theta(t)} + \frac{r(iDe^{i\phi} + (i + \omega t)Re^{i\theta(t)})}{\sqrt{D^2 + (1 + \omega^2 t^2)R^2 + 2DR(cos\alpha + \omega t sin\beta)}}] \tag{5.33}$$

$$vo_l(t) = \frac{1}{t}[De^{i\phi} + (1 - i\omega t)Re^{i\theta(t)} - \frac{r(iDe^{i\phi} + (i + \omega t)Re^{i\theta(t)})}{\sqrt{D^2 + (1 + \omega^2 t^2)R^2 + 2DR(cos\alpha + \omega t sin\beta)}}]. \tag{5.34}$$

Given $D, \phi, R, \omega, \theta_0$, the boundary of the $NLVO$ is computed simply by varying t in equations (5.30-5.34) from an initial $dt \neq 0$ (to avoid division by zero) to some preselected time horizon. The circular trajectory may represent a second order estimation of the actual trajectory followed by the obstacle.

5.5 Examples

Figure 5.6 shows three obstacles and their NLVO as seen by the point robot A. The v-obstacle of one obstacle is a straight cone since it moves at a constant

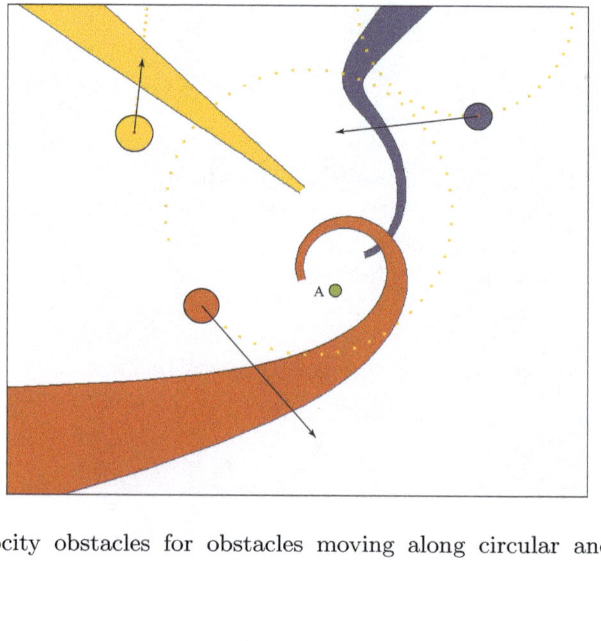

Fig. 5.6. Velocity obstacles for obstacles moving along circular and straight line trajectories

velocity. The v-obstacles of the other two are curved because of their circular trajectories. They terminate at A as time approaches infinity.

Figure 5.7 shows snapshots of an interactive planner that avoids several circular obstacles using a constant velocity [4]. The obstacles can move along straight lines or along circular trajectories at selected speeds. For the case shown, the robot A is to avoid five obstacles that move along concentric curves, and one that moves along a straight line, using a *constant* velocity from its initial position at time t_0 in frame (a) of Figure 5.7. Frame (b) shows the NLVO at time t_0, and one velocity of A that points to an opening between the many $NLVO$'s, and hence ensures avoidance of all obstacles, as shown in the remaining frames.

Figure 5.8 shows snapshots of a trajectory planner that drives the robot through a roundabout, using v-obstacles [4] [5]. To guide the robot (blue) along the road, the road boundaries are modeled by a series of circles. The roundabout is modeled by a larger circle. The planner selects the robot's speed so that it merges with the moving vehicles at their current speeds. The use of linear v-obstacles for this example would fail since the moving vehicles appear as if they are about to collide with the robot as it approaches the roundabout. This in turn can cause the robot to retreat and never reach the circular path. The nonlinear v-obstacles, on the other hand, show that no collision is imminent if the vehicles stay on their planned path, thus allowing the robot to merge with the circling traffic.

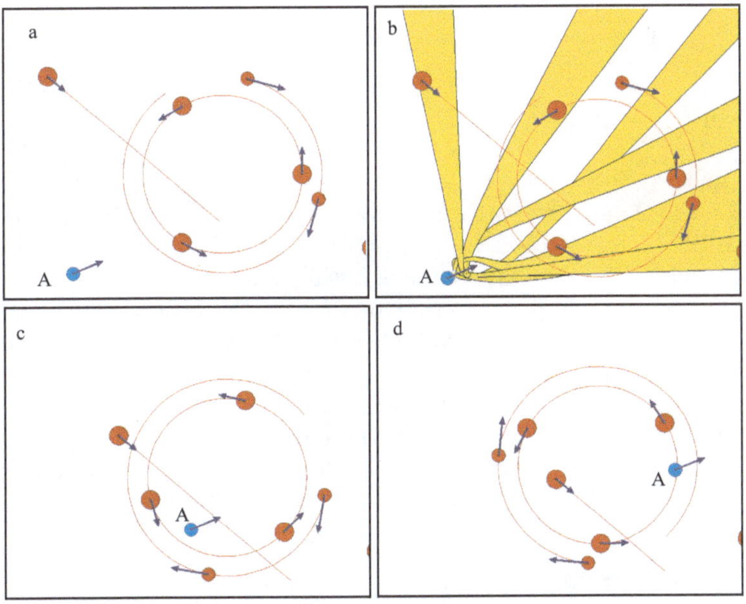

Fig. 5.7. Avoiding several moving obstacles at a constant velocity

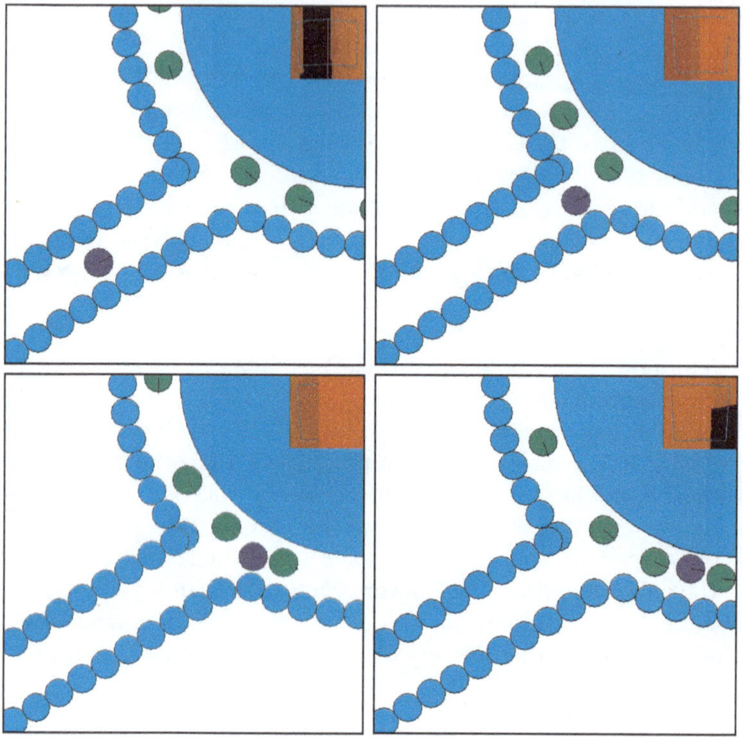

Fig. 5.8. Merging with traffic around a roundabout

5.6 Conclusions

The non-linear velocity obstacles provide a useful tool for planning motions in dynamic environments with obstacles moving along arbitrary trajectories. It consists of a warped cone that is a time-scaled map of the volume swept by the obstacle along its trajectory. Selecting a single velocity vector outside the nonlinear v-obstacle guarantees avoidance of the obstacle during the time interval for which the v-obstacle was generated. The boundaries of the non-linear v-obstacles for general planar obstacles can be computed using simple analytic expressions. They can be used to approximate unknown trajectories by using the current velocity and path curvature of the moving obstacle. Such second order approximation yields more efficient avoidance maneuvers (fewer adjustments) than the first approximation offered by the linear v-obstacle.

Acknowledgments

This work was partly supported by the IMARA *(Informatique, Mathmatiques et Automatique pour la Route Automatise)* program of INRIA on intelligent road transportation.

References

[1] A. Chakravarthy and D. Ghose. Obstacle avoidance in a dynamic environment: A collision cone approach. *IEEE Transactions on Systems, Man, and Cybernetics, Part A – Systems and Humans*, 28(5):562–574, September 1998.

[2] Prassler E., Scholz J., and P. Fiorini. Navigating a robotic wheelchair in a railway station during rush-hour. *International Journal of Robotics Research*, 18(5), May 1999.

[3] P. Fiorini and Z. Shiller. Motion planning in dynamic environments using velocity obstacles. *International Journal of Robotics Research*, 17(7):760–772, July 1998.

[4] F. Large. *Navigation Autonome D'un Robot Mobile en Environnement Dynamique et Incertain.* PhD thesis, University of Savoie, France, November 2003.

[5] Laugier C. Shiller Z. Large F. Navigation among moving obstacles using the nlvo: Principles and applications to intelligent vehicles. *Autonomous Robots*, 19(2): 159 – 171, September 2005.

[6] Shiller Z. Large F. and Sekhavat S. Motion planning in dynamic environments: Obstacle moving along arbitrary trajectories. In *IEEE International Conference of Automation and Robotics*, 2001.

[7] Z. Shiller and S. Sundar. Emergency lane-change maneuvers of autonomous vehicles. *ASME Journal of Dynamic Systems, Measurement and Control*, 120(1):37–44, March 1998.

Recursive Agent Modeling with Probabilistic Velocity Obstacles for Mobile Robot Navigation Among Humans*

Boris Kluge[1] and Erwin Prassler[2]

[1] InMach Intelligente Maschinen GmbH
Lise-Meitner-Str. 14, 89081 Ulm, Germany
kluge@inmach.de

[2] Bonn-Rhein-Sieg University of Applied Science
Grantham-Allee 20, 53757 Sankt Augustin, Germany
erwin.prassler@fh-brs.de

Summary. An approach to motion planning among moving obstacles is presented, whereby obstacles are modeled as intelligent decision-making agents. The decision-making processes of the obstacles are assumed to be similar to that of the mobile robot. A probabilistic extension to the velocity obstacle approach is used as a means for navigation and modeling uncertainty about the moving obstacles' decisions.

6.1 Introduction

Motion planning for a robot in an environment containing obstacles is an old and basic problem in robotics. For the task of navigating a mobile robot among moving obstacles, numerous approaches have been proposed. However, moving obstacles are most commonly assumed to be traveling without having any perception or motion goals (i.e. collision avoidance or goal positions) of their own.

In the expanding domain of mobile service robots deployed in natural, everyday environments, this assumption does not hold, since humans (which are the moving obstacles in this context) do perceive the robot and its motion and adapt their own motion accordingly. Therefore, reflective navigation approaches which include reasoning about other agents' navigational decision processes become increasingly interesting.

In this paper an approach to reflective navigation is presented which extends the velocity obstacle navigation scheme to incorporate reasoning about other objects' perception and motion goals.

6.1.1 Related Work

Predictive navigation is a domain where a prediction of the future motion of the obstacles is used to yield more successful motion (with respect to travel time

* This article has been published previously in E. Prassler et al. (Eds.). Advances in Human-Robot Interaction. *Springer Tracts in Advanced Robotics*, 14:89–103, 2004.

C. Laugier and R. Chatila (Eds.): Autonomous Navi. in Dyn. Environ., STAR 35, pp. 121–134, 2007.
springerlink.com

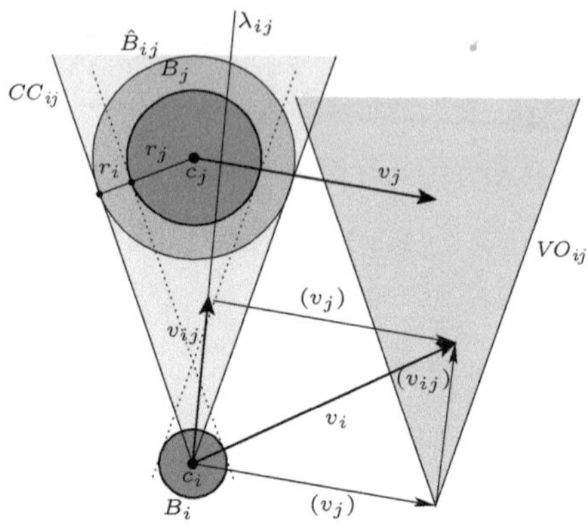

Fig. 6.1. Collision cone and velocity obstacle

or collision avoidance), see for example [3, 5]. However, reflective navigation approaches are an extension of this concept, since they include further reasoning about perception and navigational processes of moving obstacles.

The velocity obstacle paradigm, which belongs to the class of predictive navigation schemes, has been presented by Fiorini and Shiller [2] for obstacles moving on straight lines, and has been extended by Shiller et al. [6] for obstacles moving on arbitrary (but known) trajectories.

Modeling other agents' decision making similar to the own agent's decision making is used by the recursive agent modeling approach [4], where the own agent bases its decisions not only on its models of other agents' decision making processes, but also on its models of the other agents' models of its own decision making, and so on (hence the label *recursive*).

6.1.2 Overview

The remainder of this paper is organized as follows: The basic velocity obstacle approach is briefly introduced in Section 6.2, and its probabilistic extension is presented in Section 6.3. Now being able to cope with uncertain obstacle velocities, Section 6.4 describes how to recursively apply the velocity obstacle scheme in order to create a reflective navigation behavior. An implementation of the approach and experiments are given in Section 6.5. After discussing the presented work in Section 6.6, Section 6.7 concludes the paper.

6.2 Velocity Obstacles

This section gives a brief introduction to the original velocity obstacle approach [2].

Let B_i and B_j be circular objects with centers c_i and c_j and radii r_i and r_j, moving with constant velocities $v_i = \dot{c}_i$ and $v_j = \dot{c}_j$. To decide if these two objects are on a collision course, it is sufficient to consider their current positions together with their relative velocity $v_{ij} = v_i - v_j$, see Fig. 6.1. Let

$$\hat{B}_{ij} = \{c_j + r \,:\, r \in \mathbb{R}^2, |r| \leq r_i + r_j\}, \tag{6.1}$$

$$\lambda_{ij}(v_{ij}) = \{c_i + \mu v_{ij} \,:\, \mu \geq 0\}. \tag{6.2}$$

Then B_i and B_j are on a collision course, if and only if $\hat{B}_{ij} \cap \lambda_{ij}(v_{ij}) \neq \emptyset$.

Therefore we can define a set of colliding relative velocities, which is called the *collision cone* CC_{ij}, as

$$CC_{ij} = \{v_{ij} \,:\, \hat{B}_{ij} \cap \lambda_{ij}(v_{ij}) \neq \emptyset\}. \tag{6.3}$$

In order to be able to decide if an absolute velocity v_i of B_i leads to a collision with B_j, we define the *velocity obstacle* induced by B_j for B_i to be the set

$$VO_{ij} = \{v_i \,:\, (v_i - v_j) \in CC_{ij}\}, \tag{6.4}$$

which is, in other words,

$$VO_{ij} = v_j + CC_{ij}. \tag{6.5}$$

Now for B_i, any velocity $v_i \in VO_{ij}$ will lead to a collision with B_j, and any velocity $v_i \notin VO_{ij}$ for B_i will avoid any collision with B_j.

In general, object B_i is confronted with more than one other moving object. Let $\mathcal{B} = \{B_1, \ldots, B_n\}$ the set of moving objects under consideration. The *composite velocity obstacle* induced by $\mathcal{B}$ for B_i is defined as the set

$$VO_i = \cup_{j \neq i} VO_{ij}. \tag{6.6}$$

For any velocity $v_i \notin VO_i$, object B_i will not collide with any other object.

Finally, a simple navigation scheme based on velocity obstacles can be constructed as following. The moving and non-moving obstacles in the environment are continuously tracked, and the corresponding velocity obstacles are repeatedly computed. In each cycle, a reachable velocity is chosen which avoids collisions and approaches a motion goal, for example a maximum velocity towards a goal position, or a velocity minimizing the difference to a target velocity.

6.3 Probabilistic Velocity Obstacles

Let B_i and B_j be circular objects with centers c_i and c_j and radii r_i and r_j as in the previous section. However, now we will consider uncertainty in shape and velocity of the objects. This allows to reflect the limitations of real sensors and object tracking techniques.

6.3.1 Shape Uncertainty

We define the *probabilistic collision cone* of object B_j relative to object B_i to be a function

$$PCC_{ij} : \mathbb{R}^2 \to [0, 1] \tag{6.7}$$

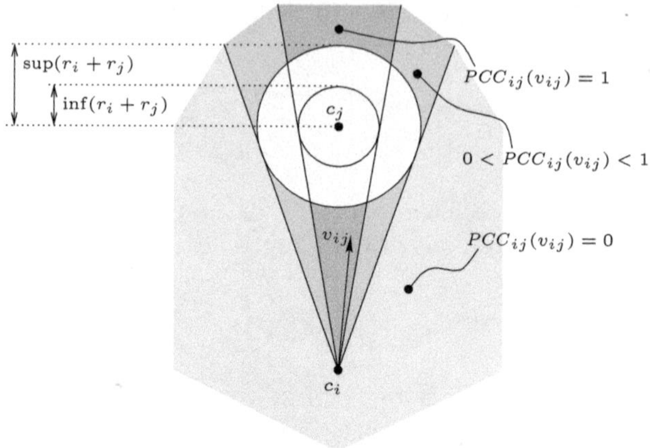

Fig. 6.2. Probabilistic collision cone

where $PCC_{ij}(v_{ij})$ is the probability of B_i to collide with B_j if B_i moves with velocity v_{ij} relative to B_j. See Fig. 6.2 for an example where a probabilistic collision cone is induced by uncertainty in the exact radii of two circular objects.

6.3.2 Velocity Uncertainty

Furthermore, there may be uncertainty about object velocities, for example introduced by a finite tracking rate. Therefore we represent the velocity of object B_j by a probability density function

$$V_j : \mathbb{R}^2 \to \mathbb{R}_0^+. \tag{6.8}$$

Now the *probabilistic velocity obstacle* of object B_j relative to object B_i is defined as the function

$$PVO_{ij} : \mathbb{R}^2 \to [0, 1] \tag{6.9}$$

which maps absolute velocities v_i of B_i to the according probability of colliding with B_j. It is

$$PVO_{ij}(v_i) = \int_{\mathbb{R}^2} V_j(v_j) PCC_{ij}(v_i - v_j) d^2 v_j \tag{6.10}$$

which is equivalent to

$$PVO_{ij} = V_j * PCC_{ij} \tag{6.11}$$

where $*$ denotes the convolution of two function. Notice the structural similarity to (6.5).

6.3.3 Composite Probabilistic Velocity Obstacle

The probability of B_i colliding with any other obstacle when traveling with velocity v_i is the probability of not avoiding collisions with any other moving obstacle. Therefore we may define the *composite probabilistic velocity obstacle* for B_i as the function

$$PVO_i = 1 - \prod_{j \neq i} (1 - PVO_{ij})\,. \tag{6.12}$$

6.3.4 Navigating with Probabilistic Velocity Obstacles

In the deterministic case, navigating is rather easy since we consider only collision free velocities and choose a velocity which is optimal for reaching the goal. But now, we are confronted with two competing objectives: reaching a goal and minimizing the probability of a collision.

Let $U_i : \mathbb{R}^2 \to [0,1]$ a function representing the utility of velocities v_i for the motion goal of B_i. Since the full utility of a velocity v_i is only attained if (a) v_i is dynamically reachable, and (b) v_i is collision free, we define the *relative utility function*

$$RU_i = U_i \cdot D_i \cdot (1 - PVO_i), \tag{6.13}$$

where $D_i : \mathbb{R}^2 \to [0,1]$ describes the reachability of a new velocity.

Now a simple navigation scheme for object B_i based on probabilistic velocity obstacles is obtained by repeatedly choosing a velocity v_i which maximizes the relative utility RU_i.

6.3.5 Implementation Problems

Objects like probabilistic velocity obstacles as presented above are continuous functions over $\mathbb{R}^2$. Therefore any implementation has to deal with a discretization of the used objects and a restriction to a finite number of cells per object.

Discretization

Step functions $s : \mathbb{R}^2 \to \mathbb{R}$ with $s(x,y) = s(x',y')$ for $i\kappa \leq x, x' < (i+1)\kappa$ and $j\kappa \leq y, y' < (j+1)\kappa$ are used for discretization in this work. In other words we use functions which are piecewise constant on squares of size $\kappa \times \kappa$, where κ is a predefined constant.

For a point $p = (x,y) \in \mathbb{R}^2$, its discretization is

$$\widehat{p} = \left(\left\lfloor \frac{x}{\kappa} \right\rfloor, \left\lfloor \frac{y}{\kappa} \right\rfloor \right) \in \mathbb{Z}^2. \tag{6.14}$$

Accordingly, for a discretized point $\widehat{p} = (z_1, z_2) \in \mathbb{Z}^2$ we define its *cell* as

$$cell(z_1, z_2) = [z_1\kappa, (z_1+1)\kappa[\times [z_2\kappa, (z_2+1)\kappa[\,. \tag{6.15}$$

For any function $F : \mathbb{R}^2 \to [0,1]$ we define the discretization of F to be the function $\widehat{F} : \mathbb{Z}^2 \to [0,1]$ with

$$\widehat{F}(z_1, z_2) = \frac{1}{\kappa^2} \int_{cell(z_1,z_2)} F(x,y)\, dx\, dy, \tag{6.16}$$

i.e. $\widehat{F}(z_1, z_2)$ is the average of F on $cell(z_1, z_2)$.

Finally, for a discretized function $\widehat{F}$,

$$\sigma(\widehat{F}) = \{(x, y) \in \mathcal{Z}^2 \mid \widehat{F}(x, y) > 0\} \tag{6.17}$$

denotes the supporting set of $\widehat{F}$, which is the set of cells on which the discretized function does not vanish. The property

$$\sigma(\widehat{F}\widehat{G}) = \sigma(\widehat{F}) \cap \sigma(\widehat{G}) \tag{6.18}$$

is easily shown. Furthermore,

$$\sum_{\widehat{p} \in \mathbb{Z}^2} \widehat{F}(\widehat{p})\kappa^2 = \int_{\mathbb{R}^2} F d^2 p \tag{6.19}$$

holds.

Restriction

We discuss this problem in the context of navigating object B_i. Assuming the velocity of any other object B_j to be bounded or to be known with bounded error, the supporting set $\sigma(\widehat{V}_j)$ is finite. Therefore, $\widehat{PVO}_{ij}(\widehat{v}_i)$ can be computed for any v_i by using

$$\widehat{PVO}_{ij}(\widehat{v}_i) = \sum_{\widehat{v}_j \in \sigma(\widehat{V}_j)} \widehat{V}_j(\widehat{v}_j)\widehat{PCC}_{ij}(\widehat{v}_i - \widehat{v}_j)\kappa^2, \tag{6.20}$$

which is the discrete version of (6.10).

Furthermore, for any real object B_i the set $\sigma(\widehat{D}_i)$ is finite, as any bounded acceleration applied to a body of non-zero mass for a bounded period of time results in a bounded change of velocity. Since only velocities from $\sigma(\widehat{RU}_i)$ will be considered for navigating B_i, and since $\sigma(\widehat{RU}_i) \subseteq \sigma(\widehat{D}_i)$, we can restrict velocities to the finite domain $\sigma(\widehat{RU}_i)$.

Algorithm

Combining the results from the previous subsections, we get

$$\widehat{PVO}_i = 1 - \prod_{j \neq i}(1 - \widehat{PVO}_{ij}) \tag{6.21}$$

and further

$$\begin{aligned} \widehat{RU}_i(\widehat{v}_i) &= \widehat{D}_i(\widehat{v}_i)\widehat{U}_i(\widehat{v}_i)(1 - \widehat{PVO}_i(\widehat{v}_i)) \\ &= \widehat{D}_i(\widehat{v}_i)\widehat{U}_i(\widehat{v}_i)\prod_{j \neq i}(1 - \widehat{PVO}_{ij}(\widehat{v}_i)) \end{aligned} \tag{6.22}$$

for any $\widehat{v}_i \in \sigma(\widehat{RU}_i)$.

Assuming that $\widehat{V}_j$, $\widehat{PCC}_{ij}$, $\widehat{D}_i$, and $\widehat{U}_i$ can be computed in time $\mathcal{O}(1)$ per function call, we can compute $\widehat{PVO}_{ij}(\widehat{v}_i)$ in time $\mathcal{O}(|\sigma(\widehat{V}_j)|)$ (according to (6.20)) and $\widehat{RU}_i(\widehat{v}_i)$ in time $\mathcal{O}(\sum_{j\neq i}|\sigma(\widehat{V}_j)|)$ (according to (6.22)). Finally, a discrete velocity $\widehat{v}_i$ maximizing $\widehat{RU}_i$ can be found in time

$$\mathcal{O}\left(|\sigma(\widehat{D}_i)| \cdot \sum_{j\neq i}|\sigma(\widehat{V}_j)|\right). \tag{6.23}$$

To put it in other words, if we assume an upper bound for $|\sigma(\widehat{V}_j)|$, the dependence of the running time on the number of obstacles is linear, but the dependence on the discretization is $\mathcal{O}(1/\kappa^4)$.

6.4 Recursive Probabilistic Velocity Obstacles

Traditionally, when navigating a mobile robot among moving obstacles (like humans), these obstacles' abilities to avoid collisions and their resulting motion behaviors are not taken into account. In contrast to this plain obstacle modeling, recursive modeling techniques presume the opponents (or more generally, the interaction partners) to deploy decision making processes for navigation similar or equal to the approach for the robot.

6.4.1 Recursive Modeling

Obstacles are assumed to perceive their environment and deduce according reactions, the reasoning process being similar to that of the robot. That is, any obstacle B_j is assumed to take actions maximizing its relative utility function RU_j, see (6.13). Therefore, in order to predict the action of obstacle B_j, we need to know its current utility function U_j, dynamic capabilities D_j, and velocity obstacle PVO_j.

The utility of velocities can be inferred by recognition of the current motion goal of the moving obstacle. For example, Bennewitz et al. [1] learn and recognize typical motion patterns of humans. If no global motion goal is available through recognition, one can still assume that there exists such a goal which the obstacle strives to approach, expecting it to be willing to keep its current speed and heading.

By continuous observation of a moving obstacle it is also possible to deduce a model of its dynamics, which describes feasible accelerations depending on its current speed and heading.

Finally, the velocity obstacle PVO_j for obstacle B_j is simply computed using the world and self models of robot B_i, assuming similar perception among the objects.

Using this information, the future motion of an obstacle can be predicted by deriving a probabilistic description of its future velocity from its relative utility function.

Now the central proposition of reflective navigation using probabilistic velocity obstacles can be expressed as

$$RU_j^d = \begin{cases} U_j D_j (1 - PVO_j^{d-1}) & \text{if } d > 0, \\ U_j D_j & \text{else,} \end{cases} \tag{6.24}$$

together with the normalization

$$V_i^d = \left(RU_i^d \right)^0 . \tag{6.25}$$

Special care has to be taken for the normalization when the involved integral vanishes, i.e. the scaling factor would become infinite. Equation (6.24) expresses that each object is assumed to deduce its relative utility from recursive probabilistic velocity obstacle considerations, while (6.25) expresses the assumption that objects will move according to their relative utility function (where applicable).

To navigate a mobile robot B_i using depth-d recursive probabilistic velocity obstacles, we repeatedly choose a velocity v_i maximizing RU_i^d. For $d = 0$, we get a behavior that only obeys the robot's utility function U_i and its dynamic capabilities D_i, but completely ignores other obstacles. For $d = 1$, we get the plain probabilistic velocity obstacle behavior as described in Section 6.3. Something new happens for $d > 1$, when the robot starts modeling the obstacles as perceptive and decision making.

6.4.2 Levels of Reflection

In the previous sections we have introduced the concept of *relative utility function* and described its recursive formulation

$$RU_i^d = D_i \cdot U_i \cdot \prod_{j \neq i} (1 - \left(RU_j^{d-1}\right)^0 * PCC_{ij}) \tag{6.26}$$

which is obtained as a combination of Equations 6.11, 6.12, 6.13, and 6.25. Reasoning about the relative utility of velocities and finding a velocity which maximizes the relative utility function on a recursion level d refers to the velocities on level $d - 1$ maximizing the relative utility functions of all other agents in the surrounding. The recursion terminates at level $d = 0$ with $RU_i^0 = D_i \cdot U_i$, where a suitable velocity is selected only based on utility and reachability. We denoted these levels of recursion as *levels of reflection.* This is motivated through the fact, that each level of recursion leads to a distinct behavior which may be associated with a certain level of contemplation or reflection.

There are basically no limits regarding the levels of recursion; at least no theoretical limits. Our results, however, showed that for pragmatic reasons we only need to consider the following four levels of recursion (and reflection).

Level $d = 0$: No Perception, No Reflection

This level does not really have a "cognitive equivalent". It is rather technically motivated: it terminates the recursion.

$$RU_i^0 = D_i \cdot U_i \tag{6.27}$$

An agent employing this level does neither perceive the environment nor does it reflect on the behavior and decision of other agents. It will ignore other agents or objects and the probabilistic velocity obstacles imposed by them and select a velocity, which maximizes only utility and reachability. Again, this level has only a technical motivation.

Level $d = 1$: Perception Without Reflection

This level in fact does have a "cognitive equivalent". When employing this level of recursion and reflection, the robot perceives the environment and the objects and other agents therein and it also perceives and accounts for their locomotion (velocity and heading) and the corresponding probabilistic obstacle velocities. On this level the relative utility function is defined as follows:

$$\begin{aligned} RU_i^1 &= D_i \cdot U_i \cdot \prod_{j \neq i} (1 - PVO_{ij}^0) \\ &= D_i \cdot U_i \cdot \prod_{j \neq i} (1 - V_j^0 * PCC_{ij}). \end{aligned} \tag{6.28}$$

The behavior shown by an agent reasoning on this level is identical with the behavior shown by an agent, which uses the plain (probabilistic) velocity obstacle without recursion.

Level $d = 2$: Reflecting About Other Agents' Intended Behavior

By employing this level of recursion the agent really reflects about the intended behavior of other agents. The formulation of its own relative utility function involves the relative utility function of its opponents at level $d = 1$:

$$\begin{aligned} RU_i^2 &= D_i \cdot U_i \cdot \prod_{j \neq i} (1 - PVO_{ij}^1) \\ &= D_i \cdot U_i \cdot \prod_{j \neq i} (1 - \left(RU_j^1\right)^0 * PCC_{ij}). \end{aligned} \tag{6.29}$$

Informally this means the robot expects the fellow agents in the surrounding environment to perceive the environment and to avoid collisions by selecting suitable velocities. So the robot in fact accounts for what he thinks that the other agents have decided to do.

An agent reasoning on this level actually shows a rather surprising behavior. Since it takes it for granted that its fellow agents can perceive the environment and avoid collisions it behaves itself in a rather ruthless and aggressive way. It pretends to accept the risk of a collision and enforces its right of way like a rude youngster with his new car (see Section Experiments).

Level $d = 3$: Reflecting About the Other Agents Reflection

This level of recursion enables the agent not only to reflect about the behavior of other agents but also about the reasoning behind their behavior. In other words, the robot reflects about the other agents reflection. The corresponding relative utility function is:

$$\begin{aligned} RU_i^3 &= D_i \cdot U_i \cdot \prod_{j \neq i} (1 - PVO_{ij}^2) \\ &= D_i \cdot U_i \cdot \prod_{j \neq i} (1 - \left(RU_j^2\right)^0 * PCC_{ij}). \end{aligned} \tag{6.30}$$

While the robot was aggressive employing level $d = 2$ it behaves rather moderate and amicable when contemplating on level $d = 3$. It assumes the other agent to show an aggressive and ruthless behavior and accordingly behaves wiser and more carefully.

As indicated above there is no technical bound for the levels of recursion and reflection, which our robot might employ in its reasoning. However, in our experiments we noticed, that for example an agent reflecting at a level $d = 4$ expects the other agents to behave wise and carefully and allows itself a more ruthless behavior showing a behaviors as if it was reasoning on level $d - 2$. Similarly, a robot employing recursion level $d = 3$ shows a behavior as if it was employing recursion level $d - 2$. Accordingly, our reflective navigation approach usually employs only the first four levels of recursion.

6.5 Implementation

Algorithm 1 describes a way to implement recursive probabilistic velocity obstacles. However not indicated for simplicity, functions are discretized according to the approach presented above.

6.5.1 Complexity

We start the complexity assessment by measuring the sizes of the supporting sets of the discretized functions used in Algorithm 1. Line 4 implies

$$\sigma(\widehat{RU}_i^{\,0}) \subseteq \sigma(\widehat{D}_i), \tag{6.31}$$

and from line 8 follows

$$\sigma(\widehat{RU}_i^{\,d}) \subseteq \sigma(\widehat{D}_i) \tag{6.32}$$

for $d > 0$. Line 3 implies

$$\sigma(\widehat{V}_i^0) = \sigma(\widehat{V}_i), \tag{6.33}$$

and from lines 11 and 13 follows

$$\sigma(\widehat{V}_i^d) \subseteq \sigma(\widehat{D}_i) \cup \sigma(\widehat{V}_i) \tag{6.34}$$

for $d > 0$, using the three preceding equations.

Algorithm 1. RPVO(DEPTH r, n OBJECTS)

1: **Input:** for $i, j = 1, \dots, n$, $j \neq i$
  - object descriptions for computing PCC_{ij}
  - velocities V_i
  - dynamic capabilities D_i
  - utilities U_i

2: **for** $i = 1, \dots, n$ **do**
3: $\quad V_i^0 \leftarrow V_i$
4: $\quad RU_i^0 \leftarrow D_i U_i$
5: **end for**
6: **for** $d = 1, \dots, r$ **do**
7: $\quad$ **for** $i = 1, \dots, n$ **do**
8: $\quad\quad RU_i^d \leftarrow D_i U_i \prod_{j \neq i} (1 - V_j^{d-1} * PCC_{ij})$
9: $\quad\quad w \leftarrow \int_{\mathbb{R}^2} RU_i^d(v) d^2 v$
10: $\quad\quad$ **if** $w > 0$ **then**
11: $\quad\quad\quad V_i^d \leftarrow (1/w) RU_i^d$
12: $\quad\quad$ **else**
13: $\quad\quad\quad V_i^d \leftarrow V_i^0$
14: $\quad\quad$ **end if**
15: $\quad$ **end for**
16: **end for**
17: **Output:** recursive models V_i^d and RU_i^d for $i = 1, \dots, n$ and $d = 0, \dots, r$

Using $N_i := |\sigma(\widehat{D}_i) \cup \sigma(\widehat{V}_i)|$, we count the numbers of operations used in the algorithm. According to (6.23), line 8 requires $\mathcal{O}(N_i \cdot \sum_{j \neq i} N_j)$ operations. Lines 9, 11, and 13 require $\mathcal{O}(N_i)$ operations each, and are thus dominated by line 8. Therefore the loop starting in line 7 requires

$$\mathcal{O}\left(\sum_{i=1}^{n}\left(N_i \sum_{j \neq i} N_j\right)\right) \tag{6.35}$$

operations, and the loop starting in line 6 requires

$$\mathcal{O}\left(r \sum_{i=1}^{n}\left(N_i \sum_{j \neq i} N_j\right)\right) \tag{6.36}$$

operations. The complexity of the loop starting in line 6 clearly dominates the complexity of the initialization loop starting in line 2. Therefore (6.36) gives an upper bound of the overall time complexity of our implementation. In other words, the dependence on the recursion depth is linear, and if we assume a common upper bound for all N_i the dependence on the number of objects is $\mathcal{O}(n^2)$, and the dependence on the discretization remains $\mathcal{O}(1/\kappa^4)$.

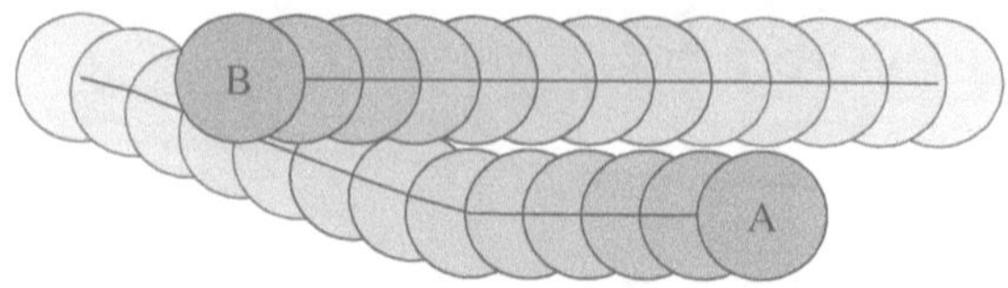

(a) Object A ($d = 2$) vs. Object B ($d = 0$)

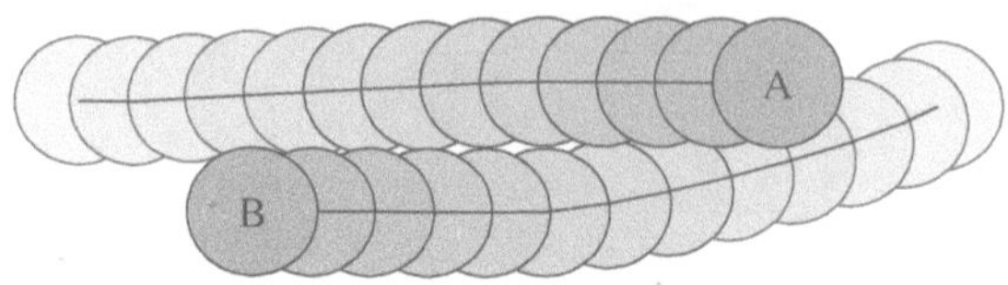

(b) Object A ($d = 2$) vs. Object B ($d = 1$)

Fig. 6.3. Collision Course Examples

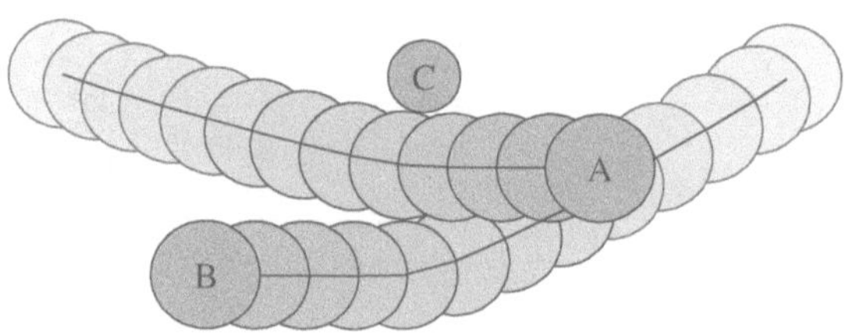

(a) Object A ($d = 2$) vs. Object B ($d = 1$)

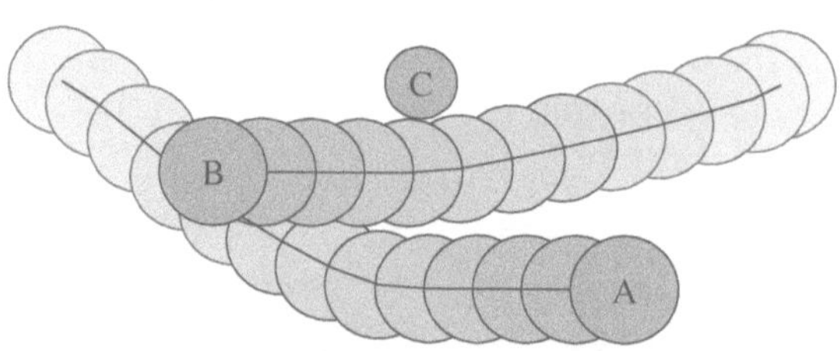

(b) Object A ($d = 3$) vs. Object B ($d = 1$)

Fig. 6.4. Static Obstacle Examples

6.5.2 Experiments

For two objects initially on an exact collision course, Fig. 6.3 shows the resulting motion for two pairs of depths. Older positions are indicated by lighter shades of grey. One can see that deeper modeling still ensures avoidance of obstacles moving obliviously (Fig. 6.3a), while collision avoiding behavior of obstacles is exploited (Fig. 6.3b).

An encounter of two objects at a static obstacle is shown in Fig. 6.4. In Fig. 6.4(a), the object from the right correctly assumes the other object to avoid collisions and exploits that anticipated behavior by choosing the shorter path around the obstacle. When reasoning to depth 3, an object assumes that the other objects expect it to avoid collisions, which results in a more defensive behavior, as it is visible in Fig. 6.4(b).

6.6 Discussion

Determining the correct evaluation depth is not obvious. If the time to collision is still large enough, we can use deeper models to possibly achieve better motion coordination. Otherwise, if obstacles and collisions are imminent, we might better give in and use a smaller depth of recursion. An alternative approach might try to combine different depth relative utility models of an object.

Similarly, finding good models of other object's dynamic capabilities D_j and utility functions U_j is a problem beyond the scope of this paper. When the action to be taken is considered the first step of a longer sequence, computing the utility function may involve motion planning, or even game-tree search, if reactions of other objects are taken into account. Due to the recursive nature of the approach, such a procedure would have to be applied for any object at any recursive level. This renders such enhancements of utility functions rather infeasible, since already single applications of such procedures are computationally expensive.

Oscillations appear in models for successive depths. Reconsider the collision course examples from Section 6.5.2 with both objects facing each other. Assume at depth d, both objects avoid a collision by deviating to the left or to the right. Then in depth $d+1$, none of the objects will perform an avoidance maneuver, since each object's depth-d model of the other object predicts that other object to avoid the collision. Subsequently, in depth $d+2$, both objects will perform collision avoidance maneuvers again, an so on.

Another aspect of the presented approach is that it can serve as a basis for reasoning about the objects in the environment. One could compare the observed motion of the objects to the motions that are predicted by recursive modeling using different utility models (for example ranging from very defensive behavior up to a homicidal chauffeur), and classify them accordingly.

Finally, it is not clear to what extent human behavior can be modeled in the presented manner. However, the approach is considered a reasonable tradeoff between computational feasibility and more detailed modeling.

6.7 Conclusion

An approach to motion coordination in dynamic environments has been presented, which reflects the peculiarities of natural, populated environments: obstacles are not only moving, but also perceiving and making decisions based on their perception.

The approach can be seen as a twofold extension of the velocity obstacle framework. Firstly, object velocities and shapes may be known and processed with respect to some uncertainty. Secondly, the perception and decision making of other objects is modeled and included in the own decision making process. So far, the approach has been tested in a simulated dynamic environment.

References

1. M. Bennewitz, W. Burgard, and S. Thrun. Learning motion patterns of persons for mobile service robots. In *Proceedings of the International Conference on Robotics and Automation (ICRA)*, 2002.
2. P. Fiorini and Z. Shiller. Motion planning in dynamic environments using velocity obstacles. *International Journal of Robotics Research*, 17(7):760–772, July 1998.
3. A. F. Foka and P. E. Trahanias. Predictive autonomous robot navigation. In *Proceedings of the 2002 IEEE/RSJ Intl. Conference on Intelligent Robots and Systems*, pages 490–495, EPFL, Lausanne, Switzerland, October 2002.
4. P. J. Gmytrasiewicz. *A Decision-Theoretic Model of Coordination and Communication in Autonomous Systems (Reasoning Systems)*. PhD thesis, University of Michigan, 1992.
5. J. Miura and Y. Shirai. Modeling motion uncertainity of moving obstacles for robot motion planning. In *Proc. of Int. Conf. on Robotics and Automation (ICRA)*, 2000.
6. Z. Shiller, F. Large, and S. Sekhavat. Motion planning in dynamic environments: Obstacles moving along arbitrary trajectories. In *Proceedings of the 2001 IEEE International Conference on Robotics and Automation*, pages 3716–3721, Seoul, Korea, May 2001.

7

Towards Real-Time Sensor-Based Path Planning in Highly Dynamic Environments

Roland Philippsen[1], Björn Jensen[2], and Roland Siegwart[3]

[1] LAAS-CNRS, Toulouse, France
roland.philippsen@gmx.net
[2] Singleton Technology, Lausanne, Switzerland
bjoern.jensen@singleton-technology.com
[3] Ecole Polytechnique Fédérale de Lausanne, Switzerland
roland.siegwart@epfl.ch

Summary. This paper presents work on sensor-based motion planning in initially unknown dynamic environments. Motion detection and probabilistic motion modeling are combined with a smooth navigation function to perform on-line path planning and replanning in cluttered dynamic environments such as public exhibitions. The SLIP algorithm, an extension of Iterative Closest Point, combines motion detection from a mobile platform with position estimation. This information is then processed using probabilistic motion prediction to yield a co-occurrence risk that unifies dynamic and static elements. The risk is translated into traversal costs for an E* path planner. It produces smooth paths that trade off collision risk versus detours.

7.1 Introduction

Path planning in a-priori unknown environments cluttered with dynamic objects is a field of active research. It can be addressed by using explicit time representation to turn the problem into an equivalent static problem, which can then be solved with an existing static planner. However, this increases the dimensionality of the representation and requires exact motion models for surrounding objects. The dimensionality increase raises the computational effort (time and memory) to produce a plan, and motion modeling raises difficult prediction issues (computationally expensive, hard to foresee the long-term evolution of real-world environments). Motion prediction becomes even more problematic in the presence of humans, and the robot is usually required to react swiftly rather than optimally. In other words: the time required to compute the plan becomes part of the optimality criterion applied to the plan.

Human behavior is unforeseeable in most situations that include human-robot interaction. As service robots or robotic companions are a highly promising application area, we actively research on-line path planning in environments with up to several hundred humans, as shown in figure 7.1. Relying on sensor-based motion modeling is crucial for correctly grounding the robot's plan. We do not try to explicitly model object and robot movements with a time-extended representation, as such complete knowledge is usually not available in the targeted

C. Laugier and R. Chatila (Eds.): Autonomous Navi. in Dyn. Environ., STAR 35, pp. 135–148, 2007.
springerlink.com

Fig. 7.1. EXPO.02 was a very crowded environment. The unpredictable behavior of humans and its dependence on the robot's actions make this a challenging real-world setting.

applications: if the humans surrounding the robot do not know where they will be going, how can the robot be expected to incorporate such knowledge during path planning?

7.1.1 Approach and Contribution

Figure 7.2 illustrates the problem statement we propose to solve with the Probabilistic Navigation Function (PNF), with the following objectives in mind:

- Avoid basing the plan on invalid assumptions. Instead, use closely sensor-based environment and object models. We identify two main reasons why models can be inappropriate: they might not correspond to the observed movements (humans rarely move with constant velocity along straight lines), or they can be hard to ground (determining model parameters by observation). The co-occurrence estimation underlying the PNF is based on few parameters that can be quickly and robustly extracted from sensory data.
- Avoid the large number of dimensions that would be required for full-fledged $\mathcal{ST}$ (state×time space) planning [5, 6]. We use worst-case scenarios to keep the dimensionality low. While this implies that no strict avoidance guarantees can be made, we have argued above that the required models would not be available for the target applications.
- Interweave planning and execution. This requires fast (re)planning and a flexible interface to motion control. By performing most computations in the workspace $\mathcal{W}$ or low-dimensional projected config uration spaces $\mathcal{C}$, planning complexity is reduced. The drawback of these speed-ups is the lack of geometrical completeness. Interfacing motion control is done via the steepest

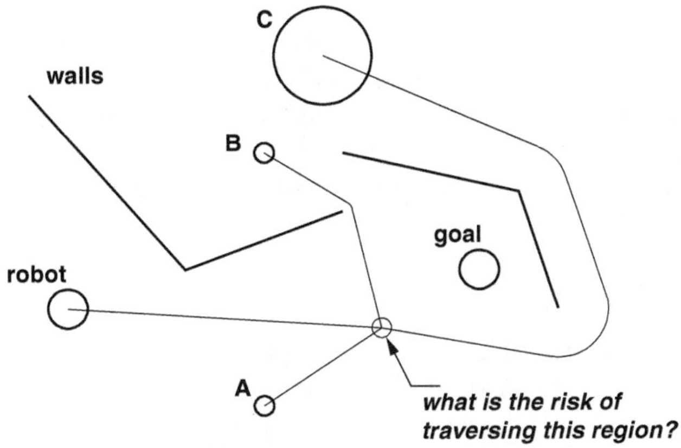

Fig. 7.2. The Probabilistic Navigation Function PNF was developed to solve the problem sketched in this diagram: *plan a path from robot to goal that strikes a balance between the accumulated risk and the detours necessary to avoid dangerous regions.* The risk is based on co-occurrence probabilities $P_{c,i}$ between the robot and each moving object. $P_{c,i}$ is estimated by taking into account static obstacles, approximated robot and object shapes, as well as maximum speeds. As indicated in this figure, the shortest admissible path from each object to the region of interest is used as worst-case estimate in order to avoid the very large number of dimensions that would be required for a $\mathcal{ST}$ space planning approach.

negative gradient of the navigation function, which is defined in (almost) all the regions the robot might reach during its movement.

This contribution presents ongoing development of the PNF approach. Prior work has resulted in the two main building blocks of motion detection and smooth navigation functions. Here we concentrate on presenting the link between the two.

To produce a smooth navigation functions (potentials with a globally unique minimum at the goal), we rely on E* [13, 15], a grid-based weighted-region planner, requiring relatively little computational overhead and being capable of dynamic replanning in near real-time[1]. This is an important property for frequently changing environment models as it limits calculations to regions that actually influence the current plan.

On-line motion detection is performed by SLIP [8], a scan alignment method. In order to use sensor-based motion modeling, it is of primary importance to compensate for the ego-motion of the robot. The main sensor used in the presented work is a laser scanner, thus motion detection requires, among other things, a robust scan alignment method. The SLIP approach is based on ICP [3, 21] (a comparison of different variants can be found in [17]) and has proven its robustness on sensory data taken during Expo.02, a six-month national exhibition

[1] `http://estar.sourceforge.net/`

which took place in Switzerland during the summer of 2002, and where the Autonomous Systems Lab was present with eleven autonomous Robox tour-guides.

The interface between these two building blocks of motion detection and path planning is developed in the following sections. It is a probabilistic approach to estimating co-occurrence probabilities between the robot and surrounding dynamic obstacles (e.g. humans, other robots), given environment constraints and assumptions based on worst-case scenarios.

7.1.2 Related Work

The PNF is a navigation function, a concept of which the NF1 [11] is a classical implementation. Navigation functions are like potential fields [9], but are guaranteed to have a unique global minimum. The PNF incorporates a continuous risk measure of traversing regions of $\mathcal{W}$, which is similar to edge costs in graph-based planners, but is conceptually closer to weighted regions.

The weighted region path planning problem is described in [12, 16], but instead of pre-determined regions we use a grid-based risk map. Among the published research that incorporates environment dynamics during path planning [1, 2, 4, 5, 6, 7, 10], most seem inappropriate for an application in highly cluttered dynamic environments: they either rely on information and computational resources that are not available for such unforeseeable settings (extending $\mathcal{C}$ to a full-fledged state-time representation [5, 6], a velocity space [4, 10], essentially off-line movement simulations [2]), or are limited to constant velocity models [7, 20]. In [1], environment dynamics are treated using worst-case scenarios that take into account the sensor capacities, but it treats all known obstacle information as static during planning.

7.2 Algorithm Overview

The components that constitute the PNF approach are shown in figure 7.3. It is based on the observation that, as static objects define the environment topology, dynamic objects can be considered traversable if it is reasonable to assume that a given object will not remain at its position once the robot has moved there. The result is an approximate plan that relies on lower level obstacle avoidance to turn it into robot movements. The PNF computes a trade-off between the collision risk of traversing a region and the detour needed if it was to be completely avoided. Accepting a certain collision risk is useful in the target application[2]. Otherwise, the robot would often not move at all or replan too frequently to be useful e.g. for tour guiding in mass exhibitions. The overall steps of the algorithm are:

Input: Laser scanner data (and odometry / localisation)
Motion detection: SLIP precisely determines the transformation between subsequent scans, then reliably detects motion that takes into account occlusion changes from a moving platform, as summarized in section 7.3.

[2] Provided that collision avoidance and other safety features of the robot perform reliably.

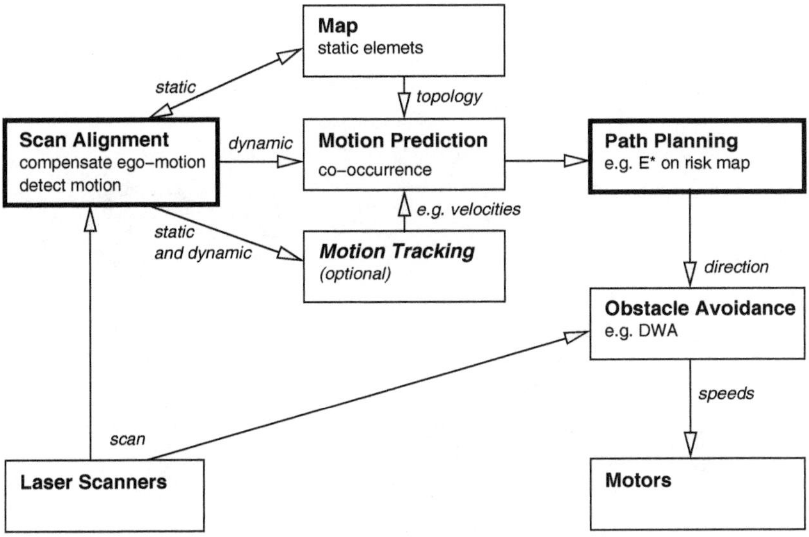

Fig. 7.3. Overview-diagram of the steps required for calculating the Probabilistic Navigation Function: laser scanner data is aligned with previous scans and separated into static and dynamic objects. Static elements are used to update the map. Dynamic elements can be optionally tracked in order to obtain more information such as velocity vectors. The static and dynamic information is used to calculate co-occurrence probabilities that define a risk map, which is transformed into a navigation function in the path planning step. Reactive obstacle avoidance is then used to execute the plan, which allows to take into account the most recent scan when determining the motor commands.

Determine co-occurrence risk: Translate dynamic objects into a probability of colliding with the given object at a given location, based on assumptions derived from worst-case scenarios. This is presented in section 7.4.
Path planning: Compute a smooth navigation function using E*, taking into account a fusion of all co-occurrence risks. This is also presented in section 7.4.
Output: Direction of travel (steepest gradient) that can be fed into reactive obstacle avoidance. This separation of path planning and execution has proven to work well during EXPO.02 [14, 18].

7.3 Scan Alignment and Motion Detection

Motion can be detected as differences between successive scans, because moving objects change sensor readings. Additionally, however, differences arise from occlusion changes due to the motion of the robot. Thus, the ego-motion has to be compensated prior to comparing scans. SLIP performs scan matching based on an initial guess from odometry, then iteratively establishes links between

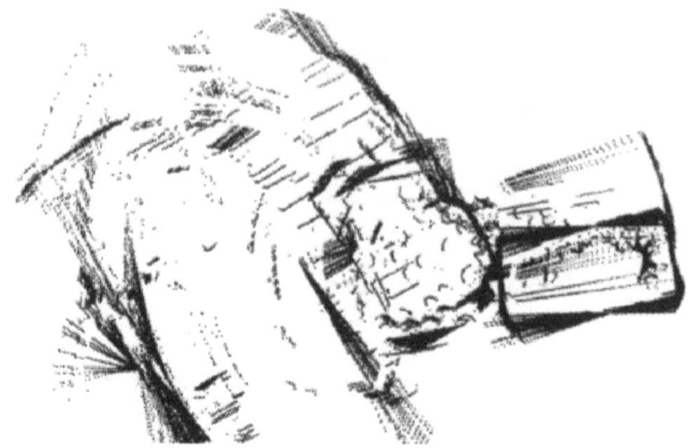

(a) scan data superimposed using raw odometry

(b) the same scans after SLIP alignment

Fig. 7.4. Example of a scan alignment: the information obtained by raw odometry is not suitable for motion detection. After correcting the ego-motion using SLIP, regions of motion are now immediately apparent. The corrected robot path is shown as a dashed line in the lower image.

points, and transforms the scans to minimise the remaining distance between the elements. Special care has to be taken to suppress outliers, particularly from moving objects, in order to achieve a high precision. Alignment correction is based on differences between the centers of gravity of the matching point sets in both scans.

To detect motion on aligned scans, elements without correspondence within a defined distance (derived from the maximal localisation error) are considered outliers. Projective filters are used to distinguish between moving objects and occlusion changes. Non-outliers are used to create a map of the static environment. SLIP then determines which outliers were visible from the previous position. An example alignment result is given in figure 7.4. Moving elements are clustered by the well known friend-of-friends algorithm to model dynamic objects with associated location (the center of gravity) and size (cluster radius). An example of motion detection from a mobile robot is shown in figure 7.5.

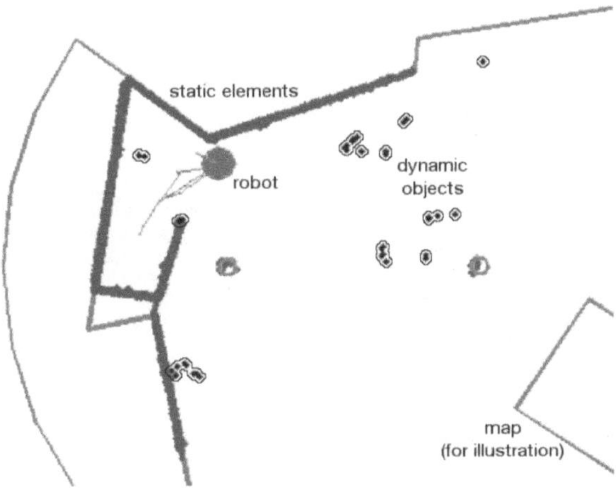

Fig. 7.5. Example of a motion detection from a mobile robot, using data acquired by the *Photobot* at EXPO.02. This was a difficult region for localisation, producing particularly large orientation errors when the robot rotated on the spot. Using SLIP, the scans were aligned very precisely to allow robust motion detection (dark spots surrounded by thin lines). Note that the map is shown for illustration only, SLIP does not require such a-priori information.

7.4 Planning with Estimated Risk

Conceptually, the co-occurrence models the probability that a given location will be occupied by a static or dynamic object by the time the robot has moved there. In principle, when all future trajectories are known, co-occurrence is a deterministic entity. However, the robot trajectory cannot be known at the planning stage, and the object movements are usually not available under real-world conditions. To cope with this, we first reduce the problem to point objects evolving in one dimension, then apply probabilistic worst-case reasoning to compute a co-occurrence estimate, and finally transform the $\mathcal{W}$-space information of non-point objects such that it can be fed into the 1D expressions.

7.4.1 One-Dimensional Co-occurrence Estimation

Instead of attempting to take into account the infinite number of trajectories that could lead to a certain region of interest $R(x)$ at a given time, we consider the case where the robot moves with v_r (as fast as possible) to $R(x)$, and then estimate the probability of the object being there as well. We can easily compute the time it takes the robot to reach any point at distance λ_r. By defining a stochastic process for the movement of the object over the distance λ_i from its current (estimated) position to $R(x)$, we can compute the probability that the object will be there when the robot arrives. We assume the object's velocity to be bounded by v_i, but that the direction of motion can change at each instant.

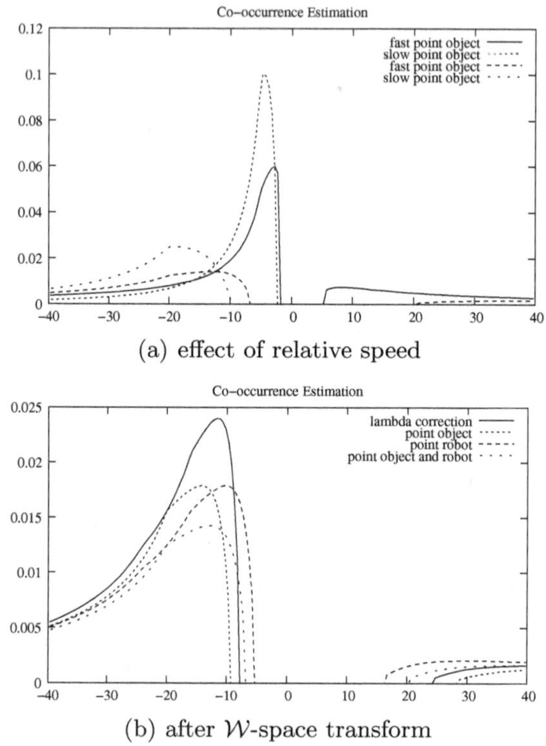

Fig. 7.6. Co-occurrence probabilities $P_{c,i}$ in the one-dimensional case. The robot is at the origin, collisions may occur on either side of the initial robot position: objects can catch up if they have higher speeds. **Top:** Four cases of dynamic object locations and speeds are shown: objects are located at $x = \{-20, -5\}$ and move with equal or twice the speed of the robot. **Bottom:** The co-occurrence is adapted to the robot and object shapes in the $\mathcal{W}$ transform, which computes the distance from a point of interest to the borders of the robot and the dynamic object. In this example, the object is located at $x = -20$ and moves twice as fast as the robot. A radius of $r_r = r_i = 4$ is used.

The robot and object are considered to be at the same location if they are within the same interval of size δ, which discretizes space into grid cells.

We start by developing the co-occurrence estimation $P_{c,i}$ in the one-dimensional case, and then we reduce locations in the workspace $\mathcal{W}$ to this 1D formulation. Figure 7.6 shows the form of probability densities that we calculate. $P_{c,i}$ (7.1) has five terms that are switched on or off depending on the robot and object speeds and reflect co-occurrences in different areas, namely far-left, left, center, right and far-right.

$$P_{c,i} = P_c(\lambda_i, \lambda_r, v_i, v_r, \delta) = p_{c,l} + p_{c,ll} + p_{c,m} + p_{c,rr} + p_{c,r} \tag{7.1}$$

where v_r is the robot speed, v_i the object speed, λ_i the distance to the object, λ_r the distance to the robot, and δ the grid resolution.

Each individual p_{c} represents an estimate for the partial co-occurrence in one of the five different areas mentioned above. Equations (7.3)-(7.7) are the expressions for these terms and the conditions under which they are non-zero. The values of $v_{1,2}$, N, and η are defined as:

$$v_{1,2} = \frac{v_{\mathrm{r}}\left(\lambda_i \mp \delta\right)}{2\lambda_{\mathrm{r}}}, \; N = \left\lceil \frac{\lambda_{\mathrm{r}}}{\delta} \right\rceil, \; \eta = \frac{N-1}{2Nv_i^2} \tag{7.2}$$

The bounds $v_{1,2}$ govern the applicability of the individual terms below. For each of the following equations, if the interval condition does not hold, the corresponding $p_{\mathrm{c}} = 0$.

$$\text{if } (v_1, v_2) \in (-\infty, -v_i) \times (-v_i, 0) \qquad p_{\mathrm{c,l}} = \frac{v_1 + v_i}{v_i} + \eta\left(v_1^2 - 2v_i^2\right) \tag{7.3}$$

$$\text{if } (v_1, v_2) \in (-v_i, 0) \times [-v_i, 0) \qquad p_{\mathrm{c,ll}} = \frac{v_2 - v_1}{v_i} + \eta\left(v_2^2 - 2v_1^2\right) \tag{7.4}$$

$$\text{if } (v_1, v_2) \in [-v_i, 0) \times [0, v_i) \qquad p_{\mathrm{c,m}} = \frac{v_2 - v_1}{v_i} - \eta\left(v_2^2 + 2v_1^2\right) \tag{7.5}$$

$$\text{if } (v_1, v_2) \in [0, v_i) \times [0, v_i) \qquad p_{\mathrm{c,rr}} = \frac{v_2 - v_1}{v_i} - \eta\left(v_2^2 - 2v_1^2\right) \tag{7.6}$$

$$\text{if } (v_1, v_2) \in [0, v_i) \times [v_i, \infty) \qquad p_{\mathrm{c,r}} = \frac{v_i - v_2}{v_i} - \eta\left(v_i^2 - 2v_2^2\right) \tag{7.7}$$

In general, the robot may travel on arbitrary collision-free paths from one point to another. Taking this into account would lead to an iterative approach of path planning and estimating the co-occurrence probabilities. But such a method is not only going to suffer from expensive computations, it may also face non-trivial convergence issues. We avoid the need for iteration by assuming that the robot will reach each point as fast as possible.

7.4.2 Risk Fusion and Planning

How can the co-occurrence information be used for path planning? The PNF is based on a combination of co-occurrence probabilities estimated in workspace $\mathcal{W}$, which is then fused and transformed to configuration space $\mathcal{C}$. The main contribution lies in the formulation of multiple layers of co-occurrence probabilities, and how these are fused into a risk map representing all dynamic objects as well as the environment topology. A flow diagram of the PNF algorithm is given in figure 7.7:

1. The distance from each grid cell to the closest static object is computed using E*. The resulting distance map is denoted D_{s}
2. D_{s} is used to compute $\mathcal{C}$-space obstacles for each dynamic object as well as for the robot. Then, similarly to the first step, E* is invoked to calculate the topologically correct distance maps D_i to the current centers of dynamic objects and D_{r} to the center of the robot.

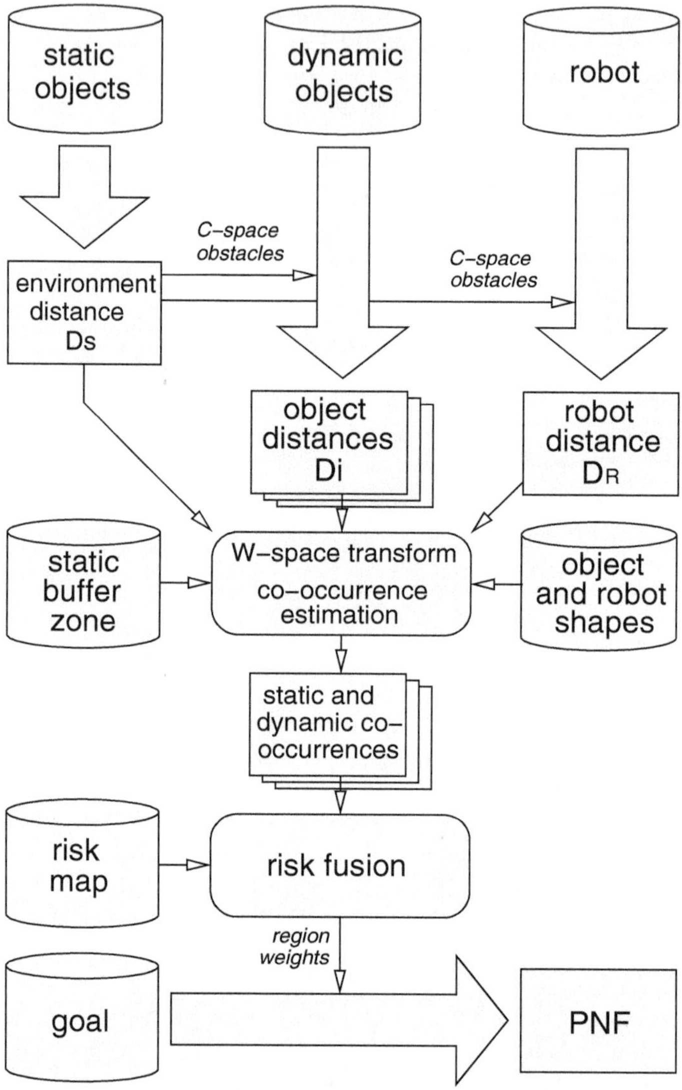

Fig. 7.7. Processing flow inside the PNF. A-priori information and inputs are denoted with cylinders, rectangles represent the various grid layers, big arrows denote E^* operation, rounded boxes are other types computations. Note that $\mathcal{C}$ obstacles are essentially binary maps of weighted regions that represent each object's and the robot's $\mathcal{C}$-space.

3. The distance maps are transformed to values that can be fed into the one-dimensional co-occurrence equations: the $\mathcal{W}$-space transform computes $\lambda_i(x) = \min_{y \in A_i(x)} (D_i(y))$, which represents the distance of a point x to the border of the dynamic object i, where $A_i(x)$ denotes the object shape placed at location x. λ_r is computed accordingly.

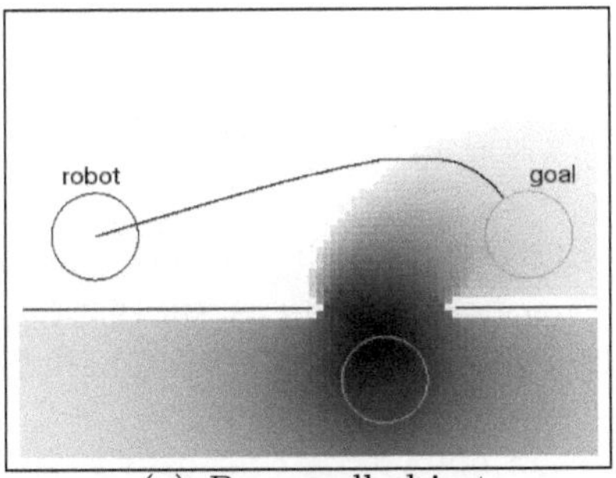

(a) $P_{c,i}$: small object

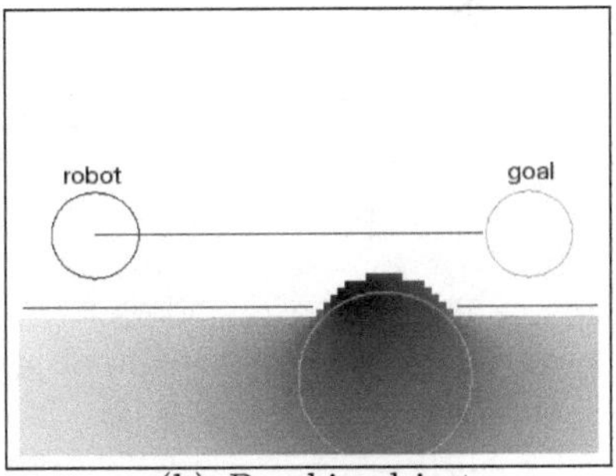

(b) $P_{c,i}$: big object

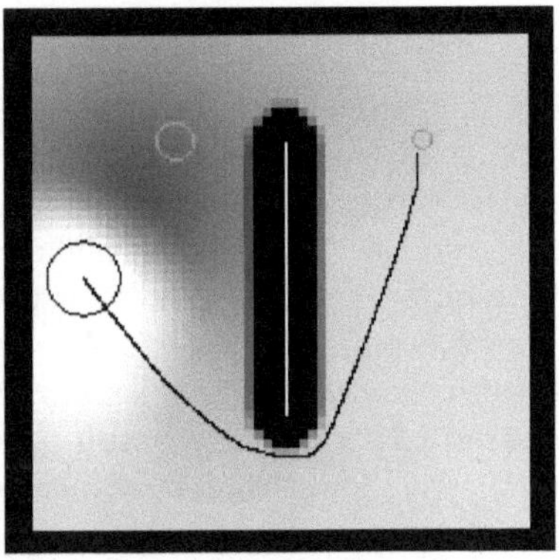

(c) P_r: $v_i = 2.5v_r$

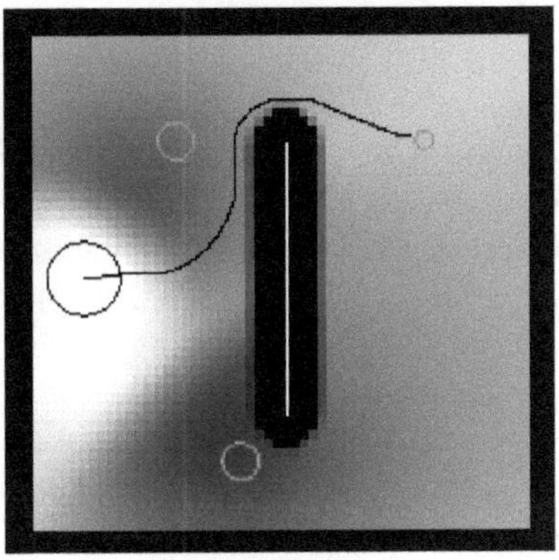

(d) P_r: $v_j = v_r$

Fig. 7.8. As shown in the two examples on the top, PNF takes into account the environment topology individually for each object: When an object is too large to fit through an opening, it cannot interfere with the robot trajectory. The two examples on the bottom illustrate how different object speeds affect the trajectory, and how the addition of a dynamic object can influence the topology of the chosen path.

4. λ_i and λ_r are fed into equation (7.1) to yield dynamic object co-occurrence maps, that is to say $P_{c,i}$ for each object at each grid cell. D_s is similarly transformed into the static co-occurrence map, after optionally applying a buffer zone which can help to smooth the robot behavior close to walls.
5. Risk fusion is performed as $P_r^{\mathcal{W}} = 1 - \prod(1 - P_c)$ over the dynamic and static co-occurrences in $\mathcal{W}$-space, followed by $P_r^{\mathcal{C}} = 1 - \prod(1 - P_r^{\mathcal{W}})$ over the robot shape to expand it to $\mathcal{C}$-space. A tunable risk map (typically of sigmoid form) is used to turn P_r into region weights.
6. Finally, E* is used again, this time to compute the navigation function, taking into account the cell weights computed with the help of the risk map. The resulting values are the Probabilistic Navigation Function.

7.5 Results

Figures 8(a) and 8(b) illustrate how PNF takes into account the environment topology for each object individually, for example in a hallway where objects can loom from rooms. The robot is the circle on the left, with a trace of gradient descent towards the goal on the right. There is a moving object behind the opening. The path is pushed into the free space in order to maximize the distance from the door, but only if the object is actually small enough to fit through. In this example, the robot speed is $v_{\mathrm{r}} = 0.3$, the object moves with $v_i = 0.2$. Figures 8(c) and 8(d) show the effect of adding an object that moves with the same speed as the robot, in this case making the path switch topology. Note the zero-risk zone around the robot (compare with figure 6(a)) and the clear line between the robot and the second object in 8(d).

The various mappings in these figures are very smooth, which is a consequence of the interpolation in E*, and could not be achieved with any strictly graph-based method. It has been shown in [13] that in the case of binary obstacle information, E* produces navigation functions very close to the true Euclidean distance (less than 10% error in typical indoor environments), whereas non-interpolating graph planners distort the distances because of the discreteness they impose on the motion choices of the robot. Interpolated dynamic replanning can be done at relatively little extra cost compared to the strictly graph-based D* [19]: the increase per operation is less than 40%, with fewer than 65% increase in the number of operations.

7.6 Conclusion and Outlook

The Probabilistic Navigation Function is an approach to on-line path planning for a-priori unknown dynamic cluttered environments. It incorporates sensor-based motion models into weighted region planning, using a probabilistic risk map based on co-occurrences. The individual building blocks were designed with on-line constraints in mind: incremental knowledge, frequent changes to environmental information, adapting existing plans, and separating planning from execution.

The finished components of the PNF are scan alignment, probabilistic collision risk estimation, and computation of the navigation function. Verifications have been carried out via simulation. Ongoing work concerns integration and testing on a real robot and implementation of higher-dimensional $\mathcal{C}$-spaces. Motion detection and ego-motion compensation were combined in the SLIP algorithm to segment sensor data into static and dynamic objects. The dynamic information is used to predict future positions, taking into account the available knowledge for each object and the static environment topology. E* is used to plan with co-occurrence information. For execution, we rely on lower level reactive obstacle avoidance guided by gradient descent, an interplay between planning and execution that has proven to perform well.

References

1. R. Alami, T. Siméon, and K. Madhava Krishna. On the influence of sensor capacities and environment dynamics onto collision-free motion plans. In *Proceedings of the IEEE/RSJ International Conference on Intelligent Robots and Systems (IROS)*, 2002.
2. Enrique J. Bernabeu, Josep Tornero, and Masayoshi Tomizuka. Collision prediction and avoidance amidst moving objects for trajectory planning applications. In *Proceedings of the IEEE International Conference on Robotics and Automation (ICRA)*, 2001.
3. P. J. Besl and N. D. Kay. A method for registration of *-D shapes. *IEEE Transactions on Pattern Analysis and Machine Intelligence*, 14(2):239–256, 1992.
4. Paolo Fiorini and Zvi Shiller. Motion planning in dynamic environments using the relative velocity paradigm. In *Proceedings of the IEEE International Conference on Robotics and Automation (ICRA)*, 1993.
5. Th. Fraichard and H. Asama. Inevitable collision states - a step towards safer robots? *Advanced Robotics*, 18(10):1001–1024, 2004.
6. Thierry Fraichard. Trajectory planning in a dynamic workspace: a 'state-time space' approach. *Advanced Robotics*, 13(1):75–94, 1999.
7. David Hsu, Robert Kindel, Jean-Claude Latombe, and Stephen Rock. Randomized kinodynamic motion planning with moving obstacles. *International Journal of Robotics Research*, 21(3):233–255, 2002.
8. Björn Jensen. *Motion Tracking for Human-Robot Interaction*. PhD thesis, Ecole Polytechnique Fédérale de Lausanne, 2004.
9. O. Khatib. Real-time obstacle avoidance for manipulators and mobile robots. *International Journal of Robotics Research*, 5(1), 1986.
10. Frederic Large, Sepanta Sekhavat, Zvi Shiller, and Christian Laugier. Towards real-time global motion planning in a dynamic environment using the NLVO concept. In *Proceedings of the IEEE/RSJ International Conference on Intelligent Robots and Systems (IROS)*, 2002.
11. J.-C. Latombe. *Robot motion planning*. Kluwer Academic Publishers, Dordrecht, Netherlands, 1991.
12. J. S. B. Mitchell and C. H. Papadimitriou. The weighted region problem: Finding shortest paths through a weighted planar subdivision. *Journal of the ACM*, 38(1):18–73, 1991.
13. Roland Philippsen. *Motion Planning and Obstacle Avoidance for Mobile Robots in Highly Cluttered Dynamic Environments*. PhD thesis, Ecole Polytechnique Fédérale de Lausanne, 2004.
14. Roland Philippsen and Roland Siegwart. Smooth and efficient obstacle avoidance for a tour guide robot. In *Proceedings of the IEEE International Conference on Robotics and Automation (ICRA)*, 2003.
15. Roland Philippsen and Roland Siegwart. An interpolated dynamic navigation function. In *Proceedings of the IEEE International Conference on Robotics and Automation (ICRA)*, 2005.
16. N. C. Rowe and R. S. Alexander. Finding optimal-path maps for path planning across weighted regions. *International Journal of Robotics Research*, 19(2):83–95, 2000.
17. S. Rusinkiewicz and M. Levoy. Efficient variants of the ICP algorithm. In *Proceedings of the Third International Conference on 3D Digital Imaging and Modeling (3DIM)*, 2001.

18. Roland Siegwart, Kai-Oliver Arras, Samir Bouabdallah, Daniel Burnier, Gilles Froidevaux, Xavier Greppin, Björn Jensen, Antoine Lorotte, Laetitia Mayor, Mathieu Meisser, Roland Philippsen, Ralph Piguet, Guy Ramel, Gregoire Terrien, and Nicola Tomatis. Robox at Expo.02: A large-scale installation of personal robots. *Robotics and Autonomous Systems*, 42:203–222, 2003.
19. Anthony Stentz. The focussed D^* algorithm for real-time replanning. In *Proceedings of the International Joint Conference on Artificial Intelligence (IJCAI)*, 1995.
20. Jur van den Berg, Dave Ferguson, and James Kuffner. Anytime path planning and replanning in dynamic environments. In *Proceedings of the IEEE International Conference on Robotics and Automation (ICRA)*, 2006.
21. Z. Zhang. Iterative point matching for registration of free-form curves and surfaces. *International Journal of Computer Vision*, 13(2):119–152, 1994.

Part III
Human-Robot Physical Interactions

8

Tasking Everyday Interaction

Elena Pacchierotti, Patric Jensfelt, and Henrik I. Christensen

Centre for Autonomous Systems, Royal Institute of Technology
SE-100 44 Stockholm, Sweden
{elenapa,patric,hic}@nada.kth.se

Summary. An important problem in the design of mobile robot systems for operation in natural environments for everyday tasks is the safe handling of encounters with people. People-People encounters follow certain social rules to allow co-existence even in cramped spaces. These social rules are often described according to the classification termed proxemics. In this paper we present an analysis of how the physical interaction with people can be modelled using the rules of proxemics and discuss how the rules of embodied feedback generation can simplify the interaction with novice users. We also provide some guidelines for the design of a control architecture for a mobile robot moving among people. The concepts presented are illustrated by a number of real experiments that verify the overall approach to the design of systems for navigation in human-populated environments.

8.1 Introduction

Robots are gradually becoming part of our daily lives. So far these robots are mostly rather simple such as autonomous vacuum cleaners. However, the fact that more than 1.000.000 such robots have been sold (Karlsson, 2004) over the past couple of years indicates that people are starting to accept the technology. There are a multitude of applications where robot technology would make a difference for the quality of life of many people, such as the assistance to the elderly and handicapped, museum tour-guidance and office assistance. Other kinds of robots are used for courier services and as part of flexible AGV-systems.

Most of these new robots have to interact with humans and in many cases also move in the same space as humans do. This puts some requirements on the robots when it comes to how they must behave. One fundamental requirement is that the robot is safe. Furthermore, there is a need to provide the robot with some basic social skills. An example of this is passage of people when encountered in the environment. When people pass each other in a corridor or on the factory floor they obey certain rules of encounter. It is natural to expect that robots, at least, should follow similar rules. This is in particular important when robots interact with users that are inexperienced or have never before met a robot. Finally the robot should not intimidate people by moving too fast or hinder people by moving too slow.

C. Laugier and R. Chatila (Eds.): Autonomous Navi. in Dyn. Environ., STAR 35, pp. 151–168, 2007.
springerlink.com

Several studies of physical interaction with people have been reported in the literature. In Nakauchi & Simmons (2000) a system that is to stand in line for event registration was reported. Here the robot has to detect the end of the line and position itself so as to obey to normal queueing behaviour. In Althaus et al. (2004) a system that participates in multi-person interaction as part of a group is described. It is here important to maintain a suitable distance from the other actors and to form a natural part of the group. Passage of people in a hallway has been reported by Yoda & Shiota (1997); an avoidance algorithm has been developed, based on a human avoidance model, where two separate conditions of a standing and walking person were considered.

In this paper we present some general guidelines for a mobile robot system that moves among people and we focus on a control strategy that is in accordance with social conventions. First, in Section 8.3, we present a control architecture that includes a layer for handling social interaction. Then, in Section 8.4, we introduce the concept of social navigation. In the rest of the paper we study the problem of social interaction of a robot with people in a hallway setting and present an algorithm for person passage that dynamically adapts the robot's behaviour to accommodate the motion of the encountered person.

8.2 Requirements

Some general requirements for the design of a robot that operates in proximity of humans can be derived from safety rules and social conventions, and are summarised below:

- The highest priority is on making sure that the system is safe so that the robot can move around humans without risking to hurt them. We believe that the requirements here are higher than they would be for a manually operated machine that performs the same task. The liability issue is central here. It will be the company that delivers the robots that takes the fall and not the operators employed by the company that use them in case of an accident.
- It is not enough for the robot to be safe, it has to be perceived as safe as well. To achieve this goal the robot must behave in a predictable and socially accepted way.
- When the robot is moving among humans it has to adapt its speed to the traffic rhythm so that it is not speeding and thereby intimidating people or moving so slowly that it hinders people.

8.3 Control Architecture Design

In mobile robotics the hybrid deliberative control paradigm (Arkin, 1998) is by now a widely adopted model for control and systems integration. The system is composed of two layers (reactive and deliberative) that compete for the control of the robot under a given policy. The reactive layer is responsible for the fast reaction to unexpected events in the environment, while the deliberative layer provides control for task achievement (see Figure 8.1). In relation to such systems, sensors provide input both to the reactive and the deliberative processes. The spatio-temporal scope of sensor information/processes determines their integration into the overall system.

No single sensor will be able to detect all obstacles around the robot and thus no single sensor will be able to guide the robot safely through a general environment. It is also unlikely that a single controller will handle all situations that the robot is faced with.

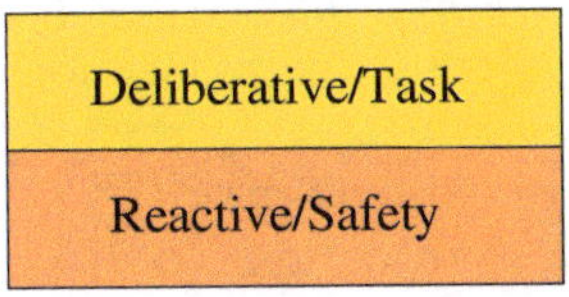

Fig. 8.1. Hybrid deliberative control architecture. The upper deliberative layer handles the global planning and the overall task management and the reactive layer works on a local scale and handles obstacle avoidance and safety.

Deliberative/Task Layer

One can distinguish between different sub-layers within the deliberative layer. At the highest level strategic planning and reasoning takes places. Lower down, closer to the reactive layer, there will be global path planning and behaviour selection. The deliberative layer stores long term information and often relies on a priori information. The bandwidth is low, from several seconds up to minutes and hours.

In most cases it is assumed that the robot has a relatively complete model of the world. The assumption is that the generated plan is going to be close to a feasible path and that the reactive layer will handle small adjustments that might be needed.

Reactive/Safety Layer

If the world was completely known, the plans generated by the deliberative layer could be executed in open loop. However, perfect knowledge is not available and unexpected events must be handled. This calls for closed loop control where information from sensors is used as input when generating the control commands. The reactive layer operates at a local scale and typically only uses a short-term memory.

In the reactive control layer we find many of the common obstacle avoidance methods (Borenstein & Koren, 1991; Fox et al., 1997; Minguez & Montano, 2004). These algorithms use a tight loop between sensors and actuators. It is well known that these methods have problems with local minima and rely on global planning from the deliberative layer for overall task achievement.

At long range the main objective is to select a path that will stay clear of obstacles and will lead to smooth motion. What is considered long distance depends on the environment and the speed of the robot. The faster the robot is moving the further ahead it needs to look to plan its path. At long distances the main requirement on the sensor is high angular resolution. Unless the angular resolution is a few degrees or better it will not be possible to segment obstacles from the background. A sonar sensor with an opening angle of 20°, for example, will not be able to detect a door opening at more than about 3 m. Currently the laser scanner offers the most robust way to detect obstacles at long range. The laser scanner is also the choice for most industrial mobile robot systems. Vision offers a more powerful source of information and it can cover more of the space. In addition it is orders of magnitude cheaper. However, to date vision algorithms do not yet offer adequate robustness.

At short range the obstacles pose a larger threat to the robot and vice versa. The main objective is therefore to be able to detect the obstacles so that they can be avoided. Ideally the sensors should cover the entire space that the robot sweeps over when it is moving, i.e., the space of possible collisions. A full sized service robot is difficult to equip with adequate sensing and it can cause significant damage. Depending on the environment some heights may be more crucial to monitor than others. For example, in most man-made indoor domestic or office environments the standard height for tables is ≈0.7 m.

8.3.1 Subdividing Reactive Layer: Emergency

In most implementations one can identify different layers within the deliberative and the reactive layers.

In Figure 8.2 we suggest a possible subdivision of the reactive layer where the lowest layer of control corresponds to a mechanism for handling emergencies.

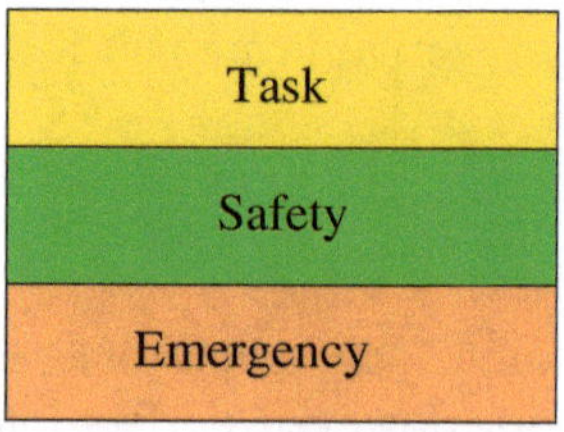

Fig. 8.2. In a more fine grained analysis of the control architecture, the reactive layer can be divided into mechanisms for handling emergency that are at the lowest level and those that deal with general safety or obstacle avoidance

The emergency layer relies on bumper type sensors. For high safety the contact sensors should be wired directly to the brakes as well as the software. This provides redundancy and also significantly increases the bandwidth and thus reduces the reaction time.

Figure 8.3 shows a situation when the non-contact sensors have not been able to detect the object in front of the robot in time resulting in a collision which is detected by the bumper sensors. The emergency control layer applies full brakes, overriding any other commands from higher up in the hierarchy.

When this lowest layer of control is activated it is a because of a failure in some of the upper layers, in this case the failure to detect the obstacle moving at a low height that is not detectable by the robot sensors. In real life these failures are often caused by not having full sensor coverage due to cost or inadequate sensors.

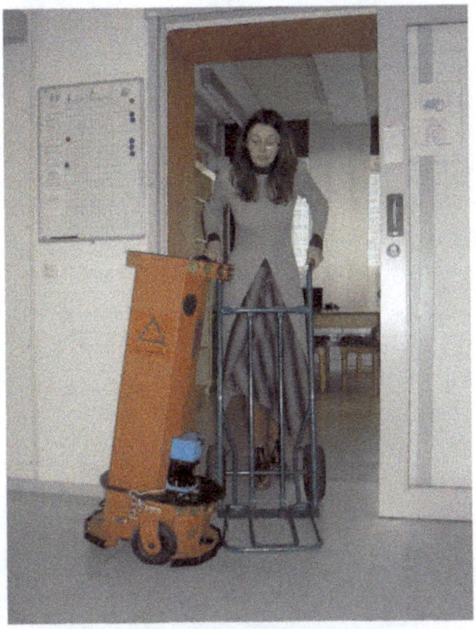

Fig. 8.3. As a last defence mechanism the robot must be able to detect collisions and activate an emergency stop to reduce harm to people or damage to property

The potential harm caused by the robot is greatly reduced if the contact sensors are complemented with non-contact sensors to reduce the speed before there is a collision. Figure 8.4 shows an example where a person steps out in front of the robot unexpectedly and the robot has to brake hard as a collision would otherwise occur. This type of functionality is part of many of the more elaborate obstacle avoidance methods as well, but the aim at the lowest level is to have a simple functionality that can be executed at a much higher rate and thus cut the reaction time and thereby reduce damage.

8.3.2 Adding Social Interaction

Guided by the requirements that the robot has to behave in a predictable and socially acceptable way, we introduce one more layer to the architecture according to Figure 8.5.

Fig. 8.4. The robot needs to monitor the environment and adapt the speed so that it is able to stop before a collision. When a collision despite this is imminent, for example when something unexpected enters the scene, the robot has to brake hard to minimize damage.

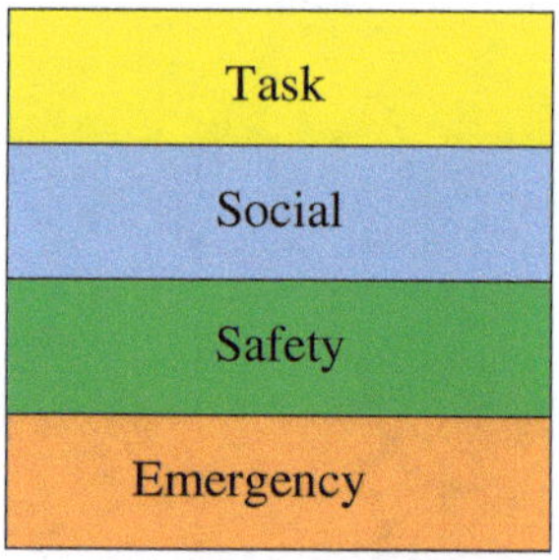

Fig. 8.5. The control architecture augmented with a layer for social interaction. This layer endows the robot with the capability of behaving in a predictable and socially accepted way.

The aim of the standard obstacle avoidance methods is to not collide with obstacles and drive the robot safely to the goal location. An objective function is often defined that takes things such as distance to obstacles, smoothness of the path and speed of travel into account. However, obstacle avoidance in general does not make sure that the robot behaves according to the social conventions as we would like. We therefore add another level on-top of the safety layer that tries to add this dimension to the behaviour of the robot when there is room for it. In Section 8.4 we will discuss this in more detail.

8.4 Social Navigation

Interaction between people is determined by normal social conventions. One such convention tells us at what side we should pass a person when we meet them in, for example, a corridor. These conventions follow the pattern of traffic. In most countries you pass a person on their left side, i.e., you walk on the right side, while in UK, Australia, Japan, ... the passage is on the other side just like in street traffic.

The interaction between people has been widely studied both in psychology and anthropology. Formal models of interaction are more recent and go back to the 1960s when the *personal space* term was defined by **?** and the *proxemics* framework was presented by Hall (1966). Good overviews on proxemics literature can be found in Aiello (1987) and Burgoon et al. (1989). In proxemics, the space around a person is divided into 4 distance zones:

Intimate distance This ranges up to 45 cm from the body and interaction within this space might include physical contact. The interaction is either directly physical such as embracing or private interaction such as whispering.

Personal distance This ranges from 0.45 m to 1.2 m and is used for interaction with family and friends or for highly organized interaction such as waiting in line.

Social distance The interaction ranges here from 1.2 m to 3.5 m and this distance is used for formal and businesslike transactions, interaction among casual acquaintances and as a separation distance in public spaces such as beaches, bus stops, shops, etc.

Public distance It extends beyond 3.5 m and is used for no interaction or in one-way interaction such as the one between an audience and a speaker.

It is important to realize that the personal space varies significantly with cultural and ethnic background. As an example, countries such as USA and the Netherlands have significantly larger distances that are to be respected in person-person interaction than Saudi Arabia and Japan.

One could model the personal space for a human as a set of elliptic regions around a person as shown in Figure 8.6. Video studies of humans in hallways seem to indicate that such a model for our spatial proxemics might be correct (Chen et al., 2004). It would be natural to assume that the robot at least should respects the same physical boundaries as we expect from other people, if the robot has to display some level of "social intelligence".

By *social navigation* we refer to a navigation strategy that takes the social aspects of interaction with people into account.

8.5 Hallway Navigation

The operation of a robot in a hallway scenario is presented here. Given that proxemics plays an important role in person-person interaction, it is of interest to study if similar rules apply for the interaction between people and robots

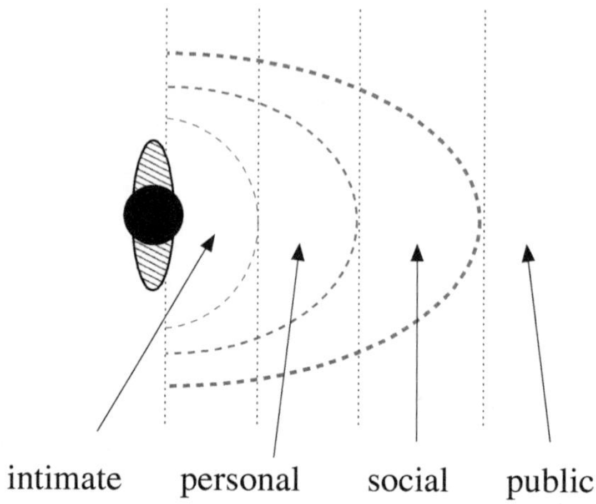

Fig. 8.6. The interaction zones for people moving through a hallway/corridor setting

operating in public spaces. Informally one would expect a robot to give way to a person when an encounter is detected. Normal human walking speed is 1-2 m/s which implies that the avoidance must be initiated early enough to signal that the robot has detected the presence of a person and to indicate its intention to provide safe passage for her/him. In the event of significant clutter the robot should move to the side of the hallway and stop until the person(s) have passed, so as to give way. A number of basic rules for the robot behaviour may thus be defined:

1. Upon entering the social space of the person initiative a move to the right (wrt. to the robot reference frame) to signal the person that has been detected.
2. Move as far to the right as the layout of the hallway allows, while passing the person.
3. Await a return to normal navigation until the person has passed by. A too early return to normal navigation might introduce discomfort on the user's side.

Using the rules of proxemics outlined in Section 8.4, one would expect the robot to initiate avoidance when the distance is about 3 meters to the person. Given a need for reliable detection, limited dynamics and early warning, however, a longer distance for reaction was chosen (6 meters). The avoidance behaviour is subject to the spatial layout of environment. If the layout is too narrow to enable passage outside of the personal space of the user, as in the case of a corridor, it is considered sufficient for the robot to move to the right as much as it is possible, respecting a safety distance from the walls. The strategy is relatively simple but at the same time it obeys the basic rules of proxemics.

8.6 An Implementation

The strategies outlined above have been implemented on a Performance PeopleBot from ActivMedia Robotics (Minnie). Minnie is equipped with a SICK laser scanner, sonar sensors and bumpers (see Figure 8.7). The system has an on board Linux computer and uses the Player/Stage software (Vaughan et al., 2003) for interfacing the robot sensors and actuators. The main components of the control system are shown in Figure 8.8.

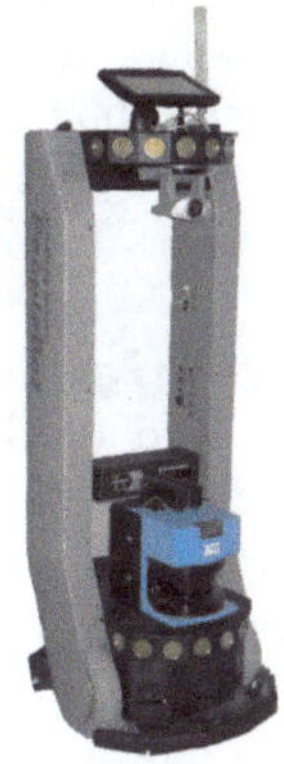

Fig. 8.7. The PeopleBot system used in our studies

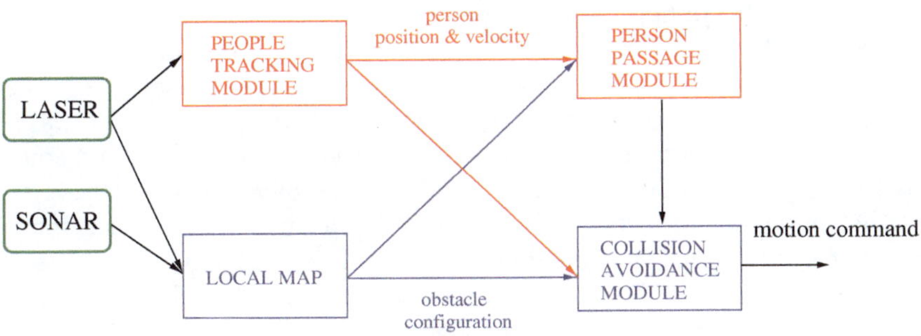

Fig. 8.8. The overall control system architecture

The laser and sonar data are fed into a local mapping system for obstacle avoidance. In addition the laser scans are fed into a person detection/tracking system. All the software runs in real-time at a rate of 10 Hz. The serial line interface to the SICK scanner runs at a rate of 5 Hz.

The tracking module detects and tracks people in the environment; the laser is mounted on the robot at a height of 33 cm from the ground to perform leg detection of the persons. Information about the current position of the people as well as their velocity is provided. Both the magnitude and the direction of the velocity are important to decide when and how to react. A particle filter, such as the one presented by Schulz et al. (2001), is used as it can deal with the presence of multiple persons.

The navigation system relies on the local mapping module that maintains a list of the closest obstacle points around the robot. Obstacle points are pruned away from the map when they are too far from the robot or when there is a closer obstacle in the same direction. The sonar data are processed through the HIMM algorithm by Borenstein & Koren (1991) before being added to the map.

The collision avoidance module can deal with significant amount of clutter but it does not take the motion of the obstacles into account as part of its planning and it does not obey the rules of social interaction. The Nearness Diagrams (ND) method by Minguez & Montano (2004) has been chosen because it is well suited for cluttered environments. The Person Passage module (PP) implements a method for navigating among dynamically changing targets and it is outlined in the next Section.

During normal operation the robot drives safely along the corridor toward an externally defined goal. The goal is feed to the collision avoidance module. In parallel the person tracker runs to detect the potential presence of a person. If a person is detected by the people tracker both the PP and the ND modules are notified. The PP module generates a strategy to pass the person. If, due to the limited width of the corridor the passage would involve entering into the personal space of the person, the ND module will override the generated motion commands and park the vehicle close the wall of the hallway, until the person has passed. Otherwise the generated motion commands are filtered through to the robot.

It is important to underline here some important assumptions that have been made in the implementation. The approach consider the presence of one person at a time; to deal with the simultaneous presence of multiple persons this strategy should be extended. It is assumed that the robot operates in a hallway wide enough to allow the simultaneous passage of the robot and the person; this means that the only impediment to the robot's maneuver is represented by the person behaviour (i.e. the person's pattern of motion along the corridor). The presented method aims at achieving a low level control modality whose only competence is to determine a passage maneuver on the right of the person, when it is possible, or to stop the robot otherwise. We believe that it is crucial to stick to this simple set of rules to avoid any ambiguity in the robot behaviour. In situations where the method decides to stop the robot, a high level module, based on a more complete set of information (localisation of the robot on a global map of the environment, user motion model for person's behaviour prediction), could determine alternative motion patterns for the robot.

8.6.1 Person Passage Method

The Person Passage module has been designed to perform a passage maneuver of a person, according to the previously defined proxemics rules. It operates as follows: as soon as a person is detected at a frontal distance below 6.0 m, the robot is steered to the right to maintain a desired lateral distance from the the user. If there is not enough space, as might be the case for a narrow corridor, the robot is commanded to move as much to the right to signal to the user that it has seen her/him and lets her/him pass.

A desired trajectory is determined that depends on the relative position and speed of the person and the environment configuration encoded in the local map. The desired trajectory is computed via a cubic spline interpolation. The control points are the current robot configuration (x_0^R, y_0^R), the desired "passage" configuration (x_P^R, y_P^R), and the final goal configuration (x^G, y^G), where x is in the direction of the corridor (see Figure 8.9).

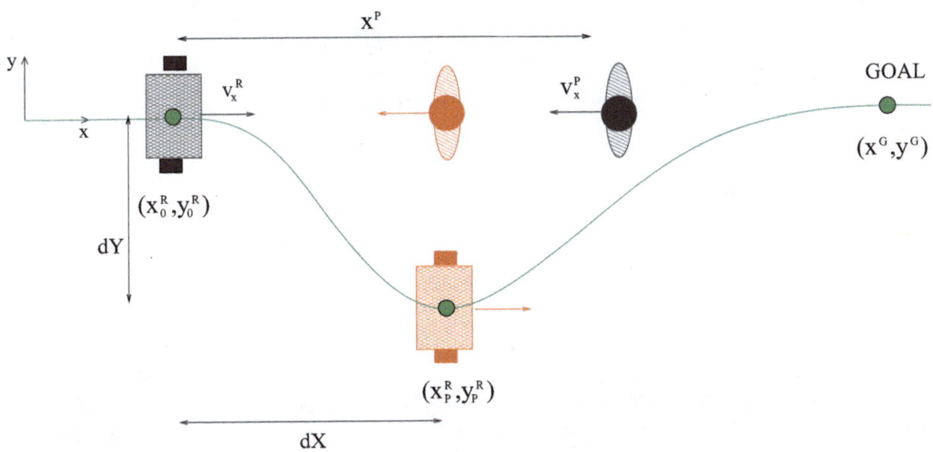

Fig. 8.9. Desired trajectory for the passage maneuver. The distance of the robot from the person is maximum when it is passing her/him (red).

The control point (x_P^R, y_P^R) determines the passage maneuver, and is computed as follows:

$$x_P^R = x_0^R + dX \tag{8.1}$$

$$y_P^R = y_0^R + dY \tag{8.2}$$

The value of dY depends on the lateral distance LD that the robot has to keep from the person:

$$dY = LD + w_R/2 - (y^P - y_0^R) \tag{8.3}$$

where w_R is the robot's width and y^P is the person's y coordinate in the corridor frame. The value of dY may be limited by the free space on the robot right. dX is computed so that the robot maintains the maximum distance from the person when it is passing her/him, according to Equation 8.4:

$$dX = v_x^R/(v_x^R - v_x^P) \times (x^P - x_0^R) \tag{8.4}$$

The robot starts the maneuver by clearly turning to the right to signal to the person its intent to pass on the right side, then the maneuver is updated according to the person's current relative position x^P and velocity v_x^P (Equation 8.4), until the person has been completely passed, at which point the robot returns to its original path. The capability to adapt to the changes in the speed of the person is crucial to establish a dynamic interaction between robot and person, as will be shown in Section 8.7, and represents an important improvement with respect to the work of Yoda & Shiota (1997).

The adopted trajectory following controller takes into account the differential drive kinematics of our robot to define the feed forward command (driving and steering velocity) (Oriolo et al., 2002):

$$v_D(t) = \sqrt{\dot{x_d}^2(t) + \dot{y_d}^2(t)} \tag{8.5}$$

$$v_S(t) = \frac{\ddot{y_d}(t)\dot{x_d}(t) - \ddot{x_d}(t)\dot{y_d}(t)}{\dot{x_d}^2(t) + \dot{y_d}^2(t)} \tag{8.6}$$

where $x_d(t)$ and $y_d(t)$ is the reference trajectory. The controller includes also an error feedback in terms of a proportional and a derivative term.

8.7 Experimental Results

The system has been evaluated in a number of different situations in the corridors of our institute, which are relatively narrow (2 m wide or less). During the experiments the "test-person" was walking at normal speed, that is around 1.0 m/s; the average speed of the robot was around 0.6 m/s.

8.7.1 Person Passage

The experiments show how the system performs in the person passage behaviour, adapting to the person speed and direction of motion. Three different cases are here presented.

In the first situation a person is walking at constant speed along the corridor. Figure 8.10 depicts top-down four different steps of the encounter. The robot starts its course in ND mode. As soon as the robot detects the person at a front distance below 6 meters, it starts its maneuver with a turn toward the right (first snapshot). This makes the person feel more comfortable and most people will instinctively move to the right too, to prepare for the passage, as it happens in the second snapshot. As soon as the person has been passed by the robot, the robot resumes its path along the center of the corridor (third and fourth frames). The steering maneuver of the robot results in an effective interaction with the user; to achieve this result, it has been crucial to perform a clear maneuver with a large advance. In the second test (see Figure 8.11), the person walks along the corridor and then stops. The robot starts its maneuver at the same front distance from the person as before (first frame) but then, detecting that the person has

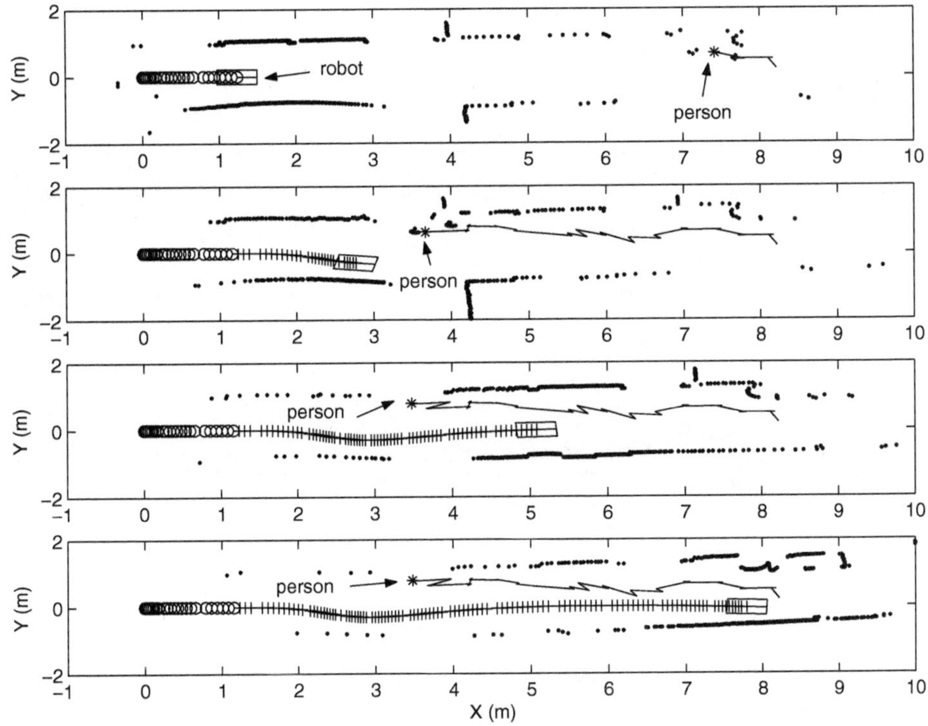

Fig. 8.10. The person walks along the corridor. The circles and the plus symbols represent the robot trajectory in ND and in PP mode, respectively. The person trajectory is shown as a continuous line and the star represents the person's current position. The current obstacle points on the local map are shown as dots. The robot steers to the right to pass the person and then resumes its course.

stopped (second frame) it does not turn toward the center of the corridor but it continues on the right until it has completely passed the person (third frame). Then the robot resumes its path toward the goal (fourth frame). Updating online the desired trajectory has allowed the robot to adapt the passage maneuver to the person relative position and velocity. This is a key feature to establish an interaction with the person that perceives the robot operation as safe and "social". In the third test (see Figure 8.12), the person is walking along the corridor and then turns to his left to enter in his office. The robot starts a maneuver of passage as before (first and second frames) but then, as soon as it detects the person on the "wrong" side of the corridor, it stops (third frame). Once the person is not detected any more, the robot resumes its path in ND mode (fourth frame). In this situation, the environment layout does not allow the robot to pass the person on the right and a passing maneuver on the left would be perceived by the person as not natural and unsafe, contradicting the

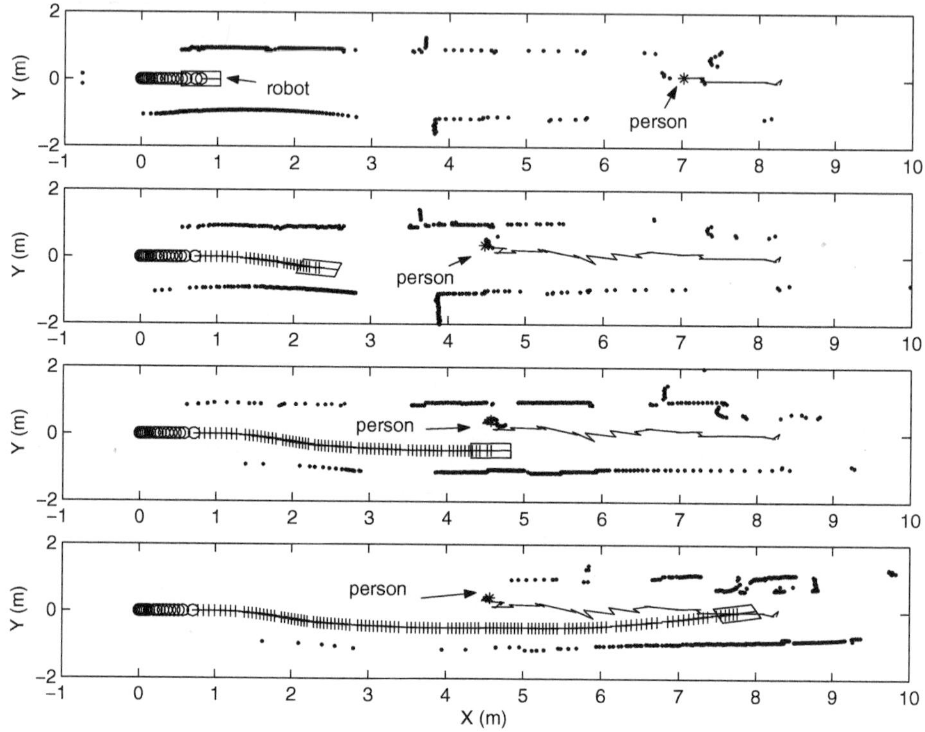

Fig. 8.11. The person stops. The robot waits until it has passed the person to resume its course on the center of the corridor.

social conventions of spatial behaviour. In such a situation, it is considered as the best solution for the robot to stop.

8.7.2 Regular Obstacles Handling

This second set of experiments show how the robot handles regular objects in the environment. A paper bin was placed in the corridor, in the robot path. The controller was in ND mode with a security distance of 0.6 m, because no persons were around. Three different configurations of the paper bin with respect to the corridor have been considered.

In the first situation the bin is on the left of the hallway, close to the wall. The robot circumvents it on the right (see Figure 8.13, left). This is automatically achieved with the ND because the right is the only free direction). It is important to observe here that ND drives the robot safely around the obstacle but it does not make the robot steer to the side as early as the PP mode does, in presence of a person. A second situation is shown on the right of Figure 8.13 in which the paper bin has been placed slightly to the right of the center of the hallway (wrt. to the robot). This is a potentially dangerous situation, because the object

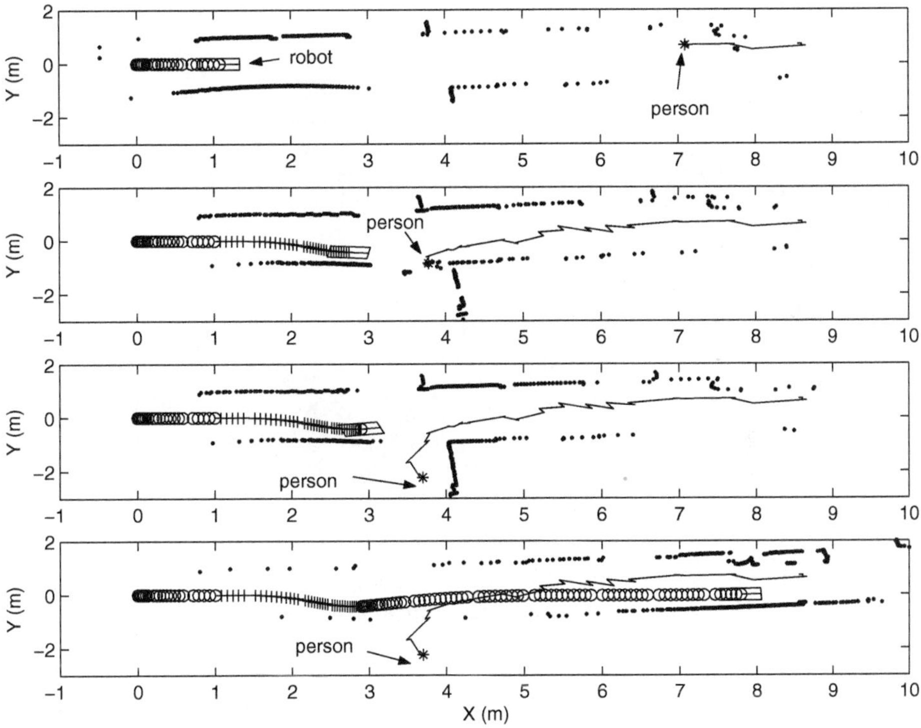

Fig. 8.12. The person crosses the robot path. The robots stops and wait until the person has disappeared from the field of view of the laser to resume its path in ND mode.

could be a non-detected person and it would be inappropriate to operate in ND mode, as ND would in most of the cases pass the obstacle on the left. The robot is not allowed to pass and it stops at a distance of 2.5 m from the object. A third case (not shown here) has been examined, where the paper bin was placed on the right side of the corridor. This is also an ambiguous situation in the case of a non-detected person, because the ND would steer the robot to the left to avoid the obstacle and this behaviour is considered not acceptable. The robot is again forced to stop at the same distance from the object as the previous case. It may appear a strong measure to stop the robot in the center of the corridor, as in the second and third situation. But it is important to underline here that we are making the assumption that the corridor should normally be free from obstacles. So, if the robot detects something in the middle of the hallway it should take into account the hypothesis that this object could be a person. In this case the chosen strategy is to stop the robot, because any other attempt to steer (as moving to the side and then stopping) could be perceived, at such short distance (2.5 m), as unsafe and unpredictable by the undetected person.

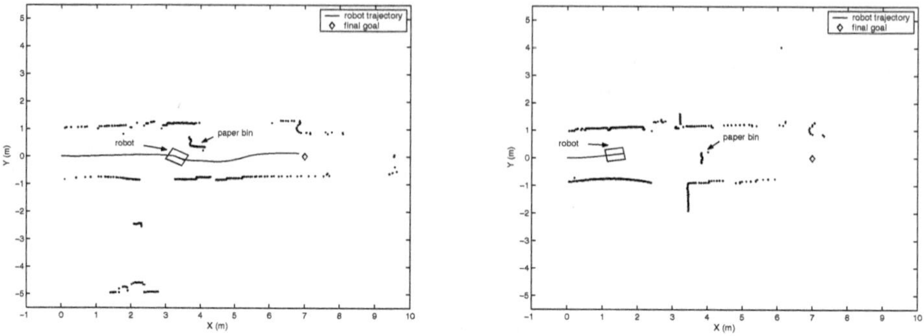

Fig. 8.13. Regular obstacles handling. On the left, the robot circumvents a paper bin placed on the left of the corridor. On the right, the paper bin is in the center of the corridor, the robot stops.

8.8 Summary/Outlook

As part of human robot interaction there is a need to consider traditional modalities such as speech, gestures and haptics, but at the same time the spatial interaction should be taken into account. For operation in environments where users might not be familiar with robots this is particularly important as it will be in general assumed that the robot behaves in a manner similar to humans. There is thus a need to transfer these rules into control laws that endow the robot with a "social" spatial behaviour.

In this paper we have presented some general guidelines for the design of a navigation system for a robot that has to move in human-populated environments. The requirements for such a robot have been formulated in terms of social interaction competences and a corresponding control architecture has been sketched.

In the rest of the paper, we have focused on the problem of passage of a person in a hallway and a control strategy has been presented, based on definitions borrowed from proxemics. The operation of the robot has been evaluated in a number of experiments in typical corridor settings which have shown how the introduction of social rules for corridor passage in the robot navigation system can give a contribution to the establishment of effective spatial interaction patterns between a robot and a person. To fully appreciate the value of such method and to fine-tune it to be socially acceptable there is a need for careful user studies. Some preliminary indications about the method have been achieved in a pilot user study in which four subjects have evaluated the acceptability of the robot motion patterns during passage with respect to three parameters: the robot speed, the signaling distance and the lateral distance kept from the person during passage (Pacchierotti et al., 2005).

The hallway passage is merely one of several different behaviours that robots must be endowed with for operation in spaces populated by people. The generalisation to other types of environments is an issue of current research.

Acknowledgements

The present research has been sponsored by the Swedish Foundation for Strategic Research (SSF) through its Centre for Autonomous Systems (CAS) and the EU as part of the Integrated Project "CoSy" (FP6-004150-IP). The support is gratefully acknowledged. H. Hüttenrauch and K. Severinson-Eklundh participation in discussions on the interaction strategy is also greatly appreciated.

References

Aiello, J. R. (1987). Human Spatial Behaviour. In D. Stokels & I. Altman (Eds.), *Handbook of Environmental Psychology.* New York, NY: John Wiley & Sons.

Althaus, P., Ishiguro, H., Kanda, T., Miyashita, T., & Christensen, H. I. (2004, April). Navigation for human-robot interaction tasks. In *Proc. of the IEEE International Conference on Robotics and Automation (icra)* (Vol. 2, p. 1894-1900).

Arkin, R. (1998). *Behavior based robotics.* Cambride, MA: AAAI Press/ The MIT Press.

Borenstein, J., & Koren, Y. (1991, Aug.). Histogramic in-motion mapping for mobile robot obstacle avoidance. *IEEE Transactions on Robotics and Automation, 7*(4), 535–539.

Burgoon, J., Buller, D., & Woodall, W. (1989). *Nonverbal Communication: The Unspoken Dialogue.* New York, NY: Harper & Row.

Chen, D., Yang, J., & Wactlar, H. D. (2004, October). Towards automatic analysis of social interaction patterns in a nursing home environment from video. In *6th ACM SIGMM International Workshop on Multimedia Information Retrieval* (Vol. Proc. of ACM MultiMedia 2004, pp. 283–290). New York, NY.

Fox, D., Burgard, W., & Thrun, S. (1997, March). The dynamic window approach to collision avoidance. *IEEE Robotics & Automation Magazine, 4* (1), 23–33.

Hall, E. T. (1966). *The hidden dimension.* New York: Doubleday.

Karlsson, J. (2004). *World robotics 2004.* Geneva, CH: United Nations Press/International Federation of Robotics.

Minguez, J., & Montano, L. (2004, Feb.). Nearness Diagram Navigation (ND): Collision avoidance in troublesome scenarios. *IEEE Transactions on Robotics and Automation, 20*(1), 45–57.

Nakauchi, Y., & Simmons, R. (2000, October). A social robot that stands in line. In *Proc. of the IEEE/RSJ International Conference on Intelligent Robots and Systems (iros)* (Vol. 1, p. 357-364).

Oriolo, G., De Luca, A., & Venditelli, M. (2002, November). WMR control via dynamic feedback linearization: design, implementation, and experimental validation. *IEEE Transactions on Control Systems Technology, 10*(6), 835–852.

Pacchierotti, E., Christensen, H. I., & Jensfelt, P. (2005, August). Human-robot embodied interaction in hallway settings: a pilot user study. In *Proc. of the IEEE Int. Workshop on Robot and Human Interactive Communication (ROMAN).* Nashville, TN.

Schulz, D., Burgard, W., Fox, D., & Cremers, A. B. (2001, December). Tracking multiple moving objects with a mobile robot. In *Proc. of the IEEE Computer Society Conference on Computer Vision and Pattern Recognition (cvpr).* Kauai, HW.

Sommer, R. (1969). *Personal Space: The Behavioral Basis of Design.* Englewood Cliffs, NJ: Prentice Hall.

Vaughan, R., Gerkey, B., & Howard, A. (2003, Oct.). On device abstraction for portable, reusable robot code. In *Proc. of the IEEE/RSJ International Conference on Intelligent Robots and Systems (IROS)* (pp. 2121–2127). Las Vegas, NV.

Yoda, M., & Shiota, Y. (1997, September). The mobile robot which passes a man. In *Proc. of the IEEE International Workshop on Robot and Human Interactive Communication (roman)* (p. 112-117).

Index

Springer Tracts in Advanced Robotics

Edited by B. Siciliano, O. Khatib and F. Groen

Vol. 35: Laugier, C.; Chatila, R. (Eds.)
Autonomous Navigation in Dynamic Environments
169 p. 2007 [978-3-540-73421-5]

Vol. 34: Wisse, M.; van der Linde, R.Q.
Delft Pneumatic Bipeds
136 p. 2007 [978-3-540-72807-8]

Vol. 33: Kong, X.; Gosselin, C.
Type Synthesis of Parallel Mechanisms
272 p. 2007 [978-3-540-71989-2]

Vol. 32: Milutinović, D.; Lima, P.
Cells and Robots: Modeling and Control of Large-Size Agent Populations
130 p. 2007 [978-3-540-71981-6]

Vol. 31: Ferre, M.; Buss, M.; Aracil, R.; Melchiorri, C.; Balaguer C. (Eds.)
Advances in Telerobotics
500 p. 2007 [978-3-540-71363-0]

Vol. 30: Brugali, D. (Ed.)
Software Engineering for Experimental Robotics
490 p. 2007 [978-3-540-68949-2]

Vol. 29: Secchi, C.; Stramigioli, S.; Fantuzzi, C.
Control of Interactive Robotic Interfaces – A Port-Hamiltonian Approach
225 p. 2007 [978-3-540-49712-7]

Vol. 28: Thrun, S.; Brooks, R.; Durrant-Whyte, H. (Eds.)
Robotics Research – Results of the 12th International Symposium ISRR
602 p. 2007 [978-3-540-48110-2]

Vol. 27: Montemerlo, M.; Thrun, S.
FastSLAM – A Scalable Method for the Simultaneous Localization and Mapping Problem in Robotics
120 p. 2007 [978-3-540-46399-3]

Vol. 26: Taylor, G.; Kleeman, L.
Visual Perception and Robotic Manipulation – 3D Object Recognition, Tracking and Hand-Eye Coordination
218 p. 2007 [978-3-540-33454-5]

Vol. 25: Corke, P.; Sukkarieh, S. (Eds.)
Field and Service Robotics – Results of the 5th International Conference
580 p. 2006 [978-3-540-33452-1]

Vol. 24: Yuta, S.; Asama, H.; Thrun, S.; Prassler, E.; Tsubouchi, T. (Eds.)
Field and Service Robotics – Recent Advances in Research and Applications
550 p. 2006 [978-3-540-32801-8]

Vol. 23: Andrade-Cetto, J,; Sanfeliu, A.
Environment Learning for Indoor Mobile Robots – A Stochastic State Estimation Approach to Simultaneous Localization and Map Building
130 p. 2006 [978-3-540-32795-0]

Vol. 22: Christensen, H.I. (Ed.)
European Robotics Symposium 2006
209 p. 2006 [978-3-540-32688-5]

Vol. 21: Ang Jr., H.; Khatib, O. (Eds.)
Experimental Robotics IX – The 9th International Symposium on Experimental Robotics
618 p. 2006 [978-3-540-28816-9]

Vol. 20: Xu, Y.; Ou, Y.
Control of Single Wheel Robots
188 p. 2005 [978-3-540-28184-9]

Vol. 19: Lefebvre, T.; Bruyninckx, H.; De Schutter, J. Nonlinear Kalman Filtering for Force-Controlled Robot Tasks
280 p. 2005 [978-3-540-28023-1]

Vol. 18: Barbagli, F.; Prattichizzo, D.; Salisbury, K. (Eds.)
Multi-point Interaction with Real and Virtual Objects
281 p. 2005 [978-3-540-26036-3]

Vol. 17: Erdmann, M.; Hsu, D.; Overmars, M.; van der Stappen, F.A (Eds.)
Algorithmic Foundations of Robotics VI
472 p. 2005 [978-3-540-25728-8]

Vol. 16: Cuesta, F.; Ollero, A.
Intelligent Mobile Robot Navigation
224 p. 2005 [978-3-540-23956-7]

Vol. 15: Dario, P.; Chatila R. (Eds.)
Robotics Research – The Eleventh International Symposium
595 p. 2005 [978-3-540-23214-8]

Vol. 14: Prassler, E.; Lawitzky, G.; Stopp, A.; Grunwald, G.; Hägele, M.; Dillmann, R.; Iossifidis. I. (Eds.)
Advances in Human-Robot Interaction
414 p. 2005 [978-3-540-23211-7]

Vol. 13: Chung, W.
Nonholonomic Manipulators
115 p. 2004 [978-3-540-22108-1]

Vol. 12: Iagnemma K.; Dubowsky, S.
Mobile Robots in Rough Terrain – Estimation, Motion Planning, and Control with Application to Planetary Rovers
123 p. 2004 [978-3-540-21968-2]

Vol. 11: Kim, J.-H.; Kim, D.-H.; Kim, Y.-J.; Seow, K.-T.
Soccer Robotics
353 p. 2004 [978-3-540-21859-3]

Vol. 10: Siciliano, B.; De Luca, A.; Melchiorri, C.; Casalino, G. (Eds.)
Advances in Control of Articulated and Mobile Robots
259 p. 2004 [978-3-540-20783-2]

Vol. 9: Yamane, K.
Simulating and Generating Motions of Human Figures
176 p. 2004 [978-3-540-20317-9]

Vol. 8: Baeten, J.; De Schutter, J.
Integrated Visual Servoing and Force Control – The Task Frame Approach
198 p. 2004 [978-3-540-40475-0]

Vol. 7: Boissonnat, J.-D.; Burdick, J.; Goldberg, K.; Hutchinson, S. (Eds.)
Algorithmic Foundations of Robotics V
577 p. 2004 [978-3-540-40476-7]

Vol. 6: Jarvis, R.A.; Zelinsky, A. (Eds.)
Robotics Research – The Tenth International Symposium
580 p. 2003 [978-3-540-00550-6]

Vol. 5: Siciliano, B.; Dario, P. (Eds.)
Experimental Robotics VIII – Proceedings of the 8th International Symposium ISER02
685 p. 2003 [978-3-540-00305-2]

Vol. 4: Bicchi, A.; Christensen, H.I.; Prattichizzo, D. (Eds.)
Control Problems in Robotics
296 p. 2003 [978-3-540-00251-2]

Vol. 3: Natale, C.
Interaction Control of Robot Manipulators – Six-degrees-of-freedom Tasks
120 p. 2003 [978-3-540-00159-1]

Vol. 2: Antonelli, G.
Underwater Robots – Motion and Force Control of Vehicle-Manipulator Systems
268 p. 2006 [978-3-540-31752-4]

Vol. 1: Caccavale, F.; Villani, L. (Eds.)
Fault Diagnosis and Fault Tolerance for Mechatronic Systems – Recent Advances
191 p. 2003 [978-3-540-44159-5]

SPRINGER NATURE

GPSR Compliance

The European Union's (EU) General Product Safety Regulation (GPSR) is a set of rules that requires consumer products to be safe and our obligations to ensure this.

If you have any concerns about our products, you can contact us on ProductSafety@springernature.com

In case Publisher is established outside the EU, the EU authorized representative is:

Springer Nature Customer Service Center GmbH
Europaplatz 3
69115 Heidelberg, Germany

Papier | Fördert gute Waldnutzung
FSC® C083411

Kolibri 360 GmbH
Ferdinand-Jühlke-Straße 7
99095 Erfurt, Deutschland
produktsicherheit@kolibri360.de